Materials for Construction: Civil Engineering Fundamentals

Materials for Construction: Civil Engineering Fundamentals

Picard Henderson

Larsen & Keller
www.larsen-keller.com

Materials for Construction: Civil Engineering Fundamentals
Picard Henderson
ISBN: 978-1-64172-115-8 (Hardback)

≣ Larsen & Keller

Published by Larsen and Keller Education,
5 Penn Plaza,
19th Floor,
New York, NY 10001, USA

Cataloging-in-Publication Data

Materials for Construction : Civil Engineering Fundamentals / Picard Henderson.
    p. cm.
Includes bibliographical references and index.
ISBN 978-1-64172-115-8
1. Building materials. 2. Civil engineering. I. Henderson, Picard.
TA403.6 .M38 2019
620.11--dc23

For more information regarding Larsen and Keller Education and its products, please visit the publisher's website www.larsen-keller.com

Table of Contents

Preface

Civil engineering is a branch of engineering that is concerned with the construction, design and maintenance of physical infrastructure such as bridges, roads, dams, airports, sewage systems, railways, etc. Many different types of building materials are used for the construction of buildings and structures. Some traditional materials are timber and blockwork. Concrete, cold rolled steel framing, composite materials, structural steel, plastics, veneer, etc. are also used in construction. This book provides comprehensive insights into the field of civil engineering. It discusses the different materials that are used in construction in extensive detail. It aims to serve as a resource guide for students and experts alike and contribute to the growth of the discipline.

A detailed account of the significant topics covered in this book is provided below:

Chapter 1, The material used for construction is termed as building material. Some common building materials are clay, sand, wood and rocks. The aim of this chapter is to provide an overview of building materials through the elucidation of topics such as cob, light straw clay, sandcrete, compressed earth block, wattle and daub, etc. **Chapter 2**, Concrete is a construction material that is composed of cement, sand and coarse aggregates that when mixed with water, hardens with time. A commonly used form of cement is the Portland cement. It is used for building beams, slabs, columns and foundations. This chapter closely examines the different types of cement and concrete used in construction, such as Gypsum concrete, precast concrete, Roman concrete, Portland cement, Geopolymer cement, fibre cement, etc. **Chapter 3**, Steel is an important construction and engineering material. It is an alloy of the elements iron, carbon, manganese, nickel, chromium and other elements in trace amounts. Wood is another common building material used for the construction of roofs, interior doors and frames, floor, etc. The topics elaborated in this chapter on structural steel, I-beam, steel channel, lumber, timber framing, etc. will help in providing an understanding of the use of steel and wood in construction. **Chapter 4**, A brick is a building material composed of clay that is used to make pavements and walls. It is of two kinds, fired and non-fired. Natural stone is used for construction. Large boulders of stone are excavated from bedrock, which are then cut and polished for creating border stones, tiles, stone blocks, etc. All the diverse topics such as cream city brick, London stock brick, Roman brick, Dutch brick, marble, limestone, etc. which are central to the use of bricks and stones in construction have been carefully examined in this chapter. **Chapter 5**, Structural dampness is one of the most common problems in construction. It is the intrusion of undesirable moisture in the structure and foundation of a building. This chapter has been carefully written to provide an understanding of the techniques of ensuring moisture protection in construction, with a close examination of the fundamental concepts of structural dampness, damp proofing, housewrap, basement waterproofing, etc.

I would like to make a special mention of my publisher who considered me worthy of this opportunity and also supported me throughout the process. I would also like to thank the editing team at the back-end who extended their help whenever required.

Picard Henderson

Common Building Materials

The material used for construction is termed as building material. Some common building materials are clay, sand, wood and rocks. The aim of this chapter is to provide an overview of building materials through the elucidation of topics such as cob, light straw clay, sandcrete, compressed earth block, wattle and daub, etc.

Building material is any material used for construction purpose such as materials for house building. Wood, cement, aggregates, metals, bricks, concrete, clay are the most common type of building material used in construction. The choices of these are based on their cost effectiveness for building projects.

Many naturally occurring substances, such as clay, sand, wood and rocks, even twigs and leaves have been used to construct buildings. Apart from naturally occurring materials, many man-made products are in use, some more and some less synthetic.

The manufacture of building materials is an established industry in many countries and the use of these materials is typically segmented into specific specialty trades, such as carpentry, plumbing, roofing and insulation work. This reference deals with habitats and structures including homes.

Types of Building Materials Used in Construction

1. Natural Construction Materials

Construction materials can be generally categorized into two sources, natural and synthetic. Natural materials are those that are unprocessed or minimally processed by industry, such as lumber or glass.

Synthetic materials are made in industrial settings after much human manipulations, such as plastics and petroleum based paints. Both have their uses.

Mud, stone, and fibrous plants are the most basic materials, aside from tents made of flexible materials such as cloth or skins. People all over the world have used these three materials together to create homes to suit their local weather conditions.

In general stone and brush are used as basic structural components in these buildings, while mud is used to fill in the space between, acting as a type of concrete and insulation.

A basic example is wattle and daub mostly used as permanent housing in tropical countries or as summer structures by ancient northern peoples.

2. Fabric

The tent used to be the home of choice among nomadic groups the world over. Two well known types include the conical teepee and the circular yurt. It has been revived as a major construction technique with the development of tensile architecture and synthetic fabrics.

Modern buildings can be made of flexible material such as fabric membranes, and supported by a system of steel cables or internal (air pressure).

3. Mud and Clay

The amount of each material used leads to different styles of buildings. The deciding factor is usually connected with the quality of the soil being used. Larger amounts of clay usually mean using the cob/adobe style, while low clay soil is usually associated with sod building.

The other main ingredients include more or less sand/gravel and straw/grasses. Rammed earth is both an old and newer take on creating walls, once made by compacting clay soils between planks by hand, now forms and mechanical pneumatic compressors are used.

Soil and especially clay is good thermal mass; it is very good at keeping temperatures at a constant level. Homes built with earth tend to be naturally cool in the summer heat and warm in cold weather. Clay holds heat or cold, releasing it over a period of time like stone.

Earthen walls change temperature slowly, so artificially raising or lowering the temperature can use more resources than in say a wood built house, but the heat/coolness stays longer.

Peoples building with mostly dirt and clay, such as cob, sod, and adobe, resulted in homes that have been built for centuries in western and northern Europe as well as the rest of the world, and continue to be built, though on a smaller scale. Some of these buildings have remained habitable for hundreds of years.

4. Rock

Rock structures have existed for as long as history can recall. It is the longest lasting building material available, and is usually readily available. There are many types of rock through out the world all with differing attributes that make them better or worse for particular uses.

Rock is a very dense material so it gives a lot of protection too, its main draw-back as a material is its weight and awkwardness. Its energy density is also considered a big draw-back, as stone is hard to keep warm without using large amounts of heating resources.

Dry-stone walls have been built for as long as humans have put one stone on top of another. Eventually different forms of mortar were used to hold the stones together, cement being the most commonplace now.

The granite-strewn uplands of Dartmoor National Park, United Kingdom, for example, provided ample resources for early settlers. Circular huts were constructed from loose granite rocks throughout the Neolithic and early Bronze Age, and the remains of an estimated 5,000 can still be seen today.

Granite continued to be used throughout the Medieval period and into modern times. Slate is another stone type, commonly used as roofing material in the United Kingdom and other parts of the world where it is found.

Mostly stone buildings can be seen in most major cities, some civilizations built entirely with stone such as the Pyramids in Egypt, the Aztec pyramids and the remains of the Inca civilization.

5. Thatch

Thatch is one of the oldest of materials known; grass is a good insulator and easily harvested. Many African tribes have lived in homes made completely of grasses year round. In Europe, thatch roofs on homes were once prevalent but the material fell out of favour as industrialization and improved transport increased the availability of other materials.

Today, though, the practice is undergoing a revival. In the Netherlands, for instance, many of new builds too have thatched roofs with special ridge tiles on top.

6. Brush

Brush structures are built entirely from plant parts and are generally found in tropical and subtropical areas, such as rainforests, where very large leaves can be used in the building. Native Americans often built brush structures for resting and living in, too.

These are built mostly with branches, twigs and leaves, and bark, similar to a beaver's lodge. These were variously named wickiups, lean-tos, and so forth.

7. Ice

Ice was used by the Inuit for igloos, but has also been used for ice hotels as a tourist attraction in northern areas that might not otherwise see many winter tourists.

8. Wood

Wood is a product of trees, and sometimes other fibrous plants, used for construction purposes when cut or pressed into lumber and timber, such as boards, planks and similar materials. It is a generic building material and is used in building just about any type of structure in most climates.

Wood can be very flexible under loads, keeping strength while bending, and is incredibly strong when compressed vertically.

There are many differing qualities to the different types of wood, even among same tree species. This means specific species are better for various uses than others. And growing conditions are important for deciding quality.

Historically, wood for building large structures was used in its unprocessed form as logs. The trees were just cut to the needed length, sometimes stripped of bark, and then notched or lashed into place.

In earlier times, and in some parts of the world, many country homes or communities had a personal wood-lot from which the family or community would grow and harvest trees to build with. These lots would be tended to like a garden.

With the invention of mechanizing saws came the mass production of dimensional lumber. This made buildings quicker to put up and more uniform. Thus the modern western style home was made.

9. Brick and Block

A brick is a block made of kiln-fired material, usually clay or shale, but also may be of lower quality mud, etc. Clay bricks are formed in a moulding (the soft mud method), or in commercial manufacture more frequently by extruding clay through a die and then wire-cutting them to the proper size (the stiff mud process).

Bricks were widely used as a construction material in the 1700, 1800 and 1900s. This was probably due to the fact that it was much more flame retardant than wood in the ever crowding cities, and fairly cheap to produce.

Another type of block replaced clay bricks in the late 20th century. It was the Cinder block. Made mostly with concrete.

An important low-cost material in developing countries is the Sandcrete block, which is weaker but cheaper than fired clay bricks.

10. Concrete

Concrete is a composite building material made from the combination of aggregate (composite) and a binder such as cement. The most common form of concrete is Portland cement concrete, which consists of mineral aggregate (generally gravel and sand), portland cement and water.

After mixing, the cement hydrates and eventually hardens into a stone-like material. When used in the generic sense, this is the material referred to by the term concrete.

For a concrete construction of any size, as concrete has a rather low tensile strength, it is generally strengthened using steel rods or bars (known as rebars). This strengthened concrete is then referred to as reinforced concrete.

In order to minimise any air bubbles, that would weaken the structure, a vibrator is used to eliminate any air that has been entrained when the liquid concrete mix is poured around the ironwork. Concrete has been the predominant material in this modern age due to its longevity, formability, and ease of transport.

11. Metal

Metal is used as structural framework for larger buildings such as skyscrapers, or as an external surface covering.

There are many types of metals used for building. Steel is a metal alloy whose major component is iron, and is the usual choice for metal structural construction. It is strong, flexible, and if refined well and/or treated lasts a long time. Corrosion is metal's prime enemy when it comes to longevity.

The lower density and better corrosion resistance of aluminium alloys and tin sometimes overcome their greater cost. Brass was more common in the past, but is usually restricted to specific uses or specialty items today.

Metal figures quite prominently in prefabricated structures such as the Quonset hut, and can be seen used in most cosmopolitan cities. It requires a great deal of human labor to produce metal, especially in the large amounts needed for the building industries.

Other metals used include titanium, chrome, gold, silver. Titanium can be used for structural purposes, but it is much more expensive than steel. Chrome, gold, and silver are used as decoration, because these materials are expensive and lack structural qualities such as tensile strength or hardness.

12. Glass

Clear windows have been used since the invention of glass to cover small openings in a building. They provided humans with the ability to both let light into rooms while at the same time keeping inclement weather outside. Glass is generally made from mixtures of sand and silicates, and is very brittle.

Modern glass "curtain walls" can be used to cover the entire facade of a building. Glass can also be used to span over a wide roof structure in a "space frame".

13. Ceramics

Ceramics are such things as tiles, fixtures, etc. Ceramics are mostly used as fixtures or coverings in buildings. Ceramic floors, walls, counter-tops, even ceilings. Many countries use ceramic roofing tiles to cover many buildings.

Ceramics used to be just a specialized form of clay-pottery firing in kilns, but it has evolved into more technical areas.

14. Plastic

The term plastic covers a range of synthetic or semi-synthetic organic condensation or polymerization products that can be molded or extruded into objects or films or fibers. Their name is derived from the fact that in their semi-liquid state they are malleable, or have the property of plasticity.

Plastics vary immensely in heat tolerance, hardness, and resiliency. Combined with this adaptability, the general uniformity of composition and lightness of plastics ensures their use in almost all industrial applications today.

15. Foam

More recently synthetic polystyrene or polyurethane foam has been used on a limited scale. It is light weight, easily shaped and an excellent insulator. It is usually used as part of a structural insulated panel where the foam is sandwiched between wood or cement.

16. Cement Composites

Cement bonded composites are an important class of construction material. These products are made of hydrated cement paste that binds wood or alike particles or fibers to make precast building components. Various fibrous materials including paper and fiberglass have been used as binders.

Wood and natural fibres are composed of various soluble organic compounds like carbohydrates, glycosides and phenolics. These compounds are known to retard cement setting. Therefore, before using a wood in making cement boned composites, its compatibility with cement is assessed.

Wood-cement compatibility is the ratio of a parameter related to the property of a wood-cement composite to that of a neat cement paste. The compatibility is often expressed as a percentage value.

To determine wood-cement compatibility, methods based on different properties are used, such as, hydration characteristics, strength, interfacial bond and morphology.

Various methods are used by researchers such as the measurement of hydration characteristics of a cement-aggregate mix; the comparison of the mechanical properties of cement-aggregate mixes and the visual assessment of microstructural properties of the wood-cement mixes.

It has been found that the hydration test by measuring the change in hydration temperature with time is the most convenient method. Recently, Karade et al. have reviewed these methods of compatibility assessment and suggested a method based on the 'maturity concept' i.e. taking in consideration both time and temperature of cement hydration reaction.

17. Building Materials in Modern Industry

Modern building is a multibillion dollar industry, and the production and harvesting of raw materials for building purposes is on a worldwide scale. Often being a primary governmental and trade keypoint between nations.

Environmental concerns are also becoming a major world topic concerning the availability and sustainability of certain materials, and the extraction of such large quantities needed for the human habitat.

18. Virtual Building Materials

Certain materials like photographs, images, text may be considered virtual. While, they usually exist on a substrate of natural material themselves, they acquire a different quality of salience to natural materials through the process of representation.

19. Building Products

When we talk about building products we refer to the ready-made particles that are fitted in different architectural hardware and decorative hardware parts of a building.

The list of building products exclusively exclude the materials, which are used to construct the building architecture and supporting fixtures like windows, doors, cabinets, etc. Building products do not make any part of a building rather they support and make them working.

Adobe

Adobe mud blocks are one of the oldest and most widely used building materials. Use of these sun-dried blocks dates back to 8000 B.C. The use of adobe is very common in some of the world's most hazard-prone regions, such as Latin America, Africa, the Indian subcontinent and other parts of Asia, the Middle East, and southern Europe. Around 30% of the world's population lives in earth-made construction. Approximately 50% of the population in developing countries, including the majority of the rural population and at least 20% of the urban and suburban population, live in earthen dwellings. By and large, mainly low-income rural populations use this type of construction.

Building in Adobe

Adobe is a low-cost, readily available construction material, usually manufactured by local communities. Typical cost of a new adobe house in Peru is about US$20/m^2 and US$11/m^2 in India. Adobe structures are generally self-made because the construction practice is simple and does not require additional energy resources. Often the blocks are made from local soil in a homeowner's yard or nearby. Mud mortar is typically used between the blocks. Skilled technicians (engineers and architects) are generally not involved in this type of construction; hence the term, "nonengineered construction," is used to describe the result.

Worldwide use of adobe is mainly in rural areas, where houses are typically one story, 3 m high, with wall thicknesses ranging from 0.25 m to 0.80 m. In mountainous regions with steep hillsides, such as the Andes, houses can be up to three stories high. In parts of the Middle East, one finds that the roof of one house is used as the floor of the house above. Urban adobe houses are found in most developing countries. However, they are not permitted by building codes in countries like Argentina, or in specific cities like San Salvador due to their poor seismic behavior.

Typical adobe house

Typical adobe house

In Latin America, adobe is mainly used by low-income families, whereas in the Middle East (e.g., Iran), it is used both by wealthy families in luxurious residences as well as by poor families in modest houses.

Architectural characteristics are similar in most countries: the rectangular plan, single door, and small lateral windows are predominant. Quality of construction in urban areas is generally superior to that in rural areas. The foundation, if present, is made of medium- to-large stones joined with mud or coarse mortar. Walls are made with adobe blocks joined with mud mortar. Sometimes straw or wheat husk is added to the soil used to make the blocks and mortar. The size of adobe blocks varies from region to region. In traditional constructions, wall thickness depends on the weather conditions of the region. Thus, in coastal areas with a mild climate, walls are thinner than in the cold highlands or in the hottest deserts. The roof is made of wood joists (usually from locally available tree trunks) resting directly on the walls or supported inside indentations on top of the walls. Roof covering may be corrugated zinc sheets or clay tiles, depending on the economic situation of the owner and the cultural inclinations of the region.

Typical adobe house in the coastal area

Typical adobe house in the highland area

A traditional adobe house that exhibits good seismic behavior is the bhonga type, typical of the Gujarat state in India. It consists of a single cylindrically shaped room with conical roof supported by cylindrical walls. It also has reinforcing bonds at the lintel and collar level, made of bamboo or reinforced concrete.

Typical adobe bhonga in India

Earthquake Performance

In addition to its low cost and simple construction technology, adobe construction has other advantages, such as excellent thermal and acoustic properties. However, most traditional adobe construction responds very poorly to earthquake ground shaking, suffering serious structural damage or collapse and causing a significant loss of life and property. In the 2001 earthquakes in El Salvador, 1,100 people died, more than 150,000 adobe buildings were severely damaged or collapsed, and over 1,600,000 people were affected. That same year, an earthquake in the south of Peru caused the deaths of 81 people, the destruction of almost 25,000 adobe houses, and damage to another 36,000 houses. In the latest 2003 Bam earthquake, more than 26,000 people died and over 60,000 were left without shelter, primarily due to the collapse of adobe houses.

Adobe buildings are not safe in seismic areas because their walls are heavy and they have low strength and brittle behavior. During strong earthquakes, due to their large mass, these structures develop high levels of seismic forces, which they are unable to resist, and therefore they fail abruptly. Typical modes of failure during earthquakes are severe cracking and disintegration of walls, separation of walls at the corners, and separation of roofs from the walls, which can lead to collapse. Seismic deficiencies characteristic of adobe construction are summarized in the figure below.

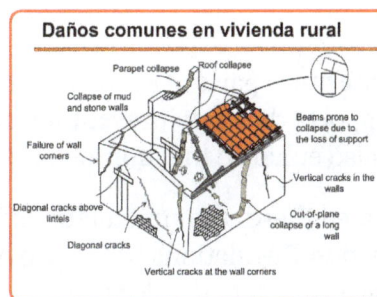

Figure: Typical modes of failure in adobe structures.

Improved Seismic Resistance of New Construction

Due to its low cost, adobe construction will continue to be used in high-risk seismic areas of the world. Development of cost-effective building technologies leading to improved seismic performance of adobe construction is of utmost importance to a substantial percentage of the global population living in adobe buildings.

The key factors for improved seismic behavior are as follows:

1.Seismic Reinforcement

The most important factor for the improved seismic performance of adobe construction is to provide reinforcement for the walls. Earthquake shaking will cause adobe walls to crack at the corners and to break up in large blocks. The role of the reinforcement therefore is to keep these large pieces of adobe wall together. A ring beam (also known as a crown, collar, bond, or tie-beam, or seismic band) that ties the walls in a box-like structure is one of the most essential components of earthquake resistance for load- bearing masonry construction. The ring beam must be strong, continuous, and well tied to the walls, and it must receive and support the roof. The ring beam can be made of concrete or timber.

Crown beam made of eucalyptus trunks   Tying adobe reinforcement

Additional wall reinforcement should also be provided. This reinforcement can be made of any strong, ductile material, such as bamboo, cane, reeds, vines, rope, timber, chicken wire, barbed wire, or steel bars. Vertical reinforcement helps to tie the wall to the foundation and to the ring beam and restrains out-of-plane bending and in-plane shear. Horizontal reinforcement helps to transmit the out-of-plane forces in transverse walls to the supporting shear walls, as well as to restrain the shear stresses between adjoining walls and to minimize vertical crack propagation. The horizontal and vertical reinforcement should be tied together and to the other structural elements to provide a stable matrix that will maintain the integrity of the walls after they have broken into large pieces.

Use of buttresses and pilasters in the critical parts of a structure increases stability and stress resistance. Buttresses act as counter supports that may prevent inward or outward overturning of the wall. Buttresses and pilasters may also enhance the interlocking of the corner bricks.

Some building codes have incorporated these recommendations for the construction of new adobe houses, such as the Adobe Construction Regulations of the province of San Juan, Argentina, that have incorporated the use of the ring beam, and the Peruvian Adobe Code that incorporated a ring beam together with vertical and horizontal reinforcement.

2. Other Considerations

The walls are the main earthquake-resisting elements of adobe houses; therefore they need to be abundant and very stable. They should be tied together to ensure mutual support and should have some reinforcement to keep them together after they have broken due to the seismic forces. The following recommendations are useful to achieve an earthquake-resistant adobe house:

- Build only one-story houses

- Use an insulated lightweight roof instead of a heavy roof (clay tiles or compacted earth)

- Keep the openings in the walls small and well spaced

- Build on firm soil and provide a concrete or stone foundation

The soil used to fabricate the adobe blocks and the mud mortar must contain clay, since it provides strength to the dry materials. Unfortunately, clay shrinks during drying; therefore an excessive amount of clay will cause cracking of the blocks and mortar due to shrinkage, and loss of strength in the adobe masonry. Straw, wheat husk, and to a lesser extent, coarse sand can be used as additives to control this cracking and thus to improve the strength of adobe masonry. The quality of workmanship also plays an important role in obtaining strong adobe masonry. Good workmanship can improve the strength of adobe masonry by up to 100%.

Seismic Strengthening of Existing Adobe Buildings

Dynamic tests conducted by researchers in Peru have demonstrated that a good solution for existing adobe houses is an external reinforcement consisting of wide strips of wire mesh (1 mm wires spaced at ¾ inches) nailed with metallic bottle caps against the adobe as shown in the figure below. The mesh is placed in horizontal and vertical strips simulating beams and columns and is covered with cement and sand mortar. Several houses reinforced with this technique did not suffer any damage during the 2001 earthquake in the south of Peru, even though similar unreinforced houses in the vicinity collapsed or suffered significant damage.

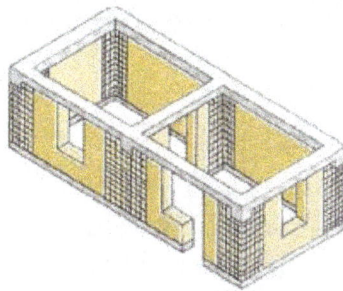

Layout of the wire mesh reinforcement

Historic adobe buildings, regardless of their important architectural or cultural value, are also prone to suffer damage during strong earthquakes. Thus, it is important to provide adequate upgrading to these buildings to ensure life-safety protection and at the same time to preserve their authenticity. The Getty Conservation Institute recently carried out a project to develop technical procedures to prevent the structural instability of historic adobe buildings during earthquakes, with minimal intervention to their original fabric. Nine small-scale (1:5) and two large-scale (1:2) model buildings were subjected to shaking table tests to compare different reinforcement systems. An effective retrofit system was developed, consisting of straps made of woven nylon placed horizontally or vertically, forming loops around the entire building or around individual walls. Nylon cross-ties were added to hold these straps. Vertical steel rods drilled directly into the adobe walls were effective in delaying and limiting both the inplane and out-of-plane wall damage. Wood bond-beams or partial wood diaphragms were placed to achieve integral participation of the walls.

Cob

Cob is a mixture of sandy-sub soil, clay and straw. It is mixed by crushing the particles together by either dancing on it or using the head of a digger. Historically cob might have been mixed by farm animals that would walk up and down on the sand, clay and straw. The sandy sub-soil must be sharp and ideally contain angular stones and gravel – this will make it stronger. About 75% of cob is made up of this sandy aggregate. Any type of clay can be used, but be careful not to use silt which can sometimes appear like clay.

The oldest cob house still standing is 10,000 years old. Cob is strong, durable and cob houses should stand forever as long as their roof is maintained and the property is looked after properly. In the UK we ensure our cob houses last for hundreds of years by incorporating a few basic design features that make them suitable for our climate. These include: a gravel foundation to stop the capillary action of water, a 50 cm stone or brick stem wall to keep the cob off the ground, a roof that overhangs by about 50 cm and a lime render on the external walls.

Cob is 75% sandy sub-soil

Cob is so easy to work with – it is a lot more forgiving and less precise than working with bricks. Of course you have to learn to keep your walls exactly verticle but if you find your wall is slightly out you simply shave off some cob with an old saw or add a bit more cob. Fitting windows and doors is so easy too – you can add more cob around the frame if your gap is too big, or chip bits away from the cob if you hole is too small. On our cob house building courses we show you how easy it is to fit your windows, doors and even roof.

Advantages of Using Cob

Cob is the most sustainable form of building there is. The materials for your cob walls are usually excavated from your foundation trench and on-site. This means there is no manufacture or transportation of materials. Many so called 'eco homes' claim to be green because they are cheap to run once built but the materials used to create them, usually have a massive carbon footprint. In contrast cob is genuinely as 'eco' as you can get as it has almost zero embodied energy. Since cob is made of the earth it is also entirely recyclable and non-polluting.

Cob can be mixed by foot

- Cob is affordable and Cob walls almost costs nothing. As long as you have some land to build on, anyone can afford to build cob walls. Our cob expert Kate Edwards was originally taught by Ianto Evans who built his own cob house in the USA for only a few hundred dollars. And we just built our own cob-bale home for about a tenth of the cost of a conventional home.

- Cob houses are breathable and healthy to live in. There is no damp in a cob house.

- Cob houses require almost no heating. On our courses we teach you how to design your cob house on passive solar principles. We show you how to use cob on the south facing walls to make the most of its excellent thermal mass, and straw bales on the north walls to provide excellent insulation. A single cob fireplace home heats our cob-bale as we used these passive solar principles in our build.

- Cob houses are beautiful. When you create a cob house you literally sculpt it. Cob allows you to easily create curves, and carve shelves and features into your walls. Of course cob houses don't have to be curvy though, they can be made with straight walls and right angles for that modern look.

Light Straw Clay

Light straw clay can be infilled in nearly every wall framing system, be it timber framing, pole framing, conventional lumber framing, or framing specifically designed for straw clay infill.

2×4 interior walls are infilled with LSC to aid in sound dampening and creating a sense of privacy for this small 800-square foot three bedroom home.

LSC is also excellent retrofit insulation because preexisting walls can be furred out to any thickness. Furring out a wall simply involves adding stud material to the desired depth of wall. This can be done to the interior of a building or to the exterior. Using staggered studs or Larsen trusses also improves the insulation's performance because it allows the creation of a continuous thermal envelope, doing away with the thermal bridging that occurs in a conventionally framed building (where solid studs create breaks between insulated stud cavities).

Interior walls can be infilled with straw clay in buildings that have exterior wall systems of other materials. Interior walls can benefit from the soundproofing that straw clay provides, and they provide a seamless look because they take plaster as well as other natural wall systems. If done with good and consistent formwork and with attention to detail, the walls can be very flat, lending themselves to very smooth finish plaster, which leads to less "dusting" through the life of the wall.

LSC's compatibility with conventional framing systems makes it easier to find contractors who can provide straightforward estimates for a project.

Wall systems or walls with lots of openings, like the south side of a passive solar building in the Northern Hemisphere, are highly compatible with straw clay, whereas cob, adobe, and straw bale are hard to work with around windows, doors, and other openings. It's a somewhat common practice to design a building that takes advantage of the high R-value of straw bales for the north, west, and east walls of a building, but use LSC in the south wall, which has the bulk of the glazing.

One of the advantages light straw clay has over cob and adobe and other natural wall systems is that it slumps and sags very little while being installed, allowing an entire wall cavity to be filled in one work session. As long as tamping is consistent and there are not long periods of drying time between installations in the same wall/ stud cavity, there is also very little shrinkage.

Many projects, particularly in urban areas, have to be carried out in limited space. When there is limited square footage to work with, the 18″ to 24″ width of straw bales or cob may rule these wall systems out because they eat into usable space. In urban areas and sites with limited space, straw clay can be an excellent choice to create thinner wall systems that are still highly insulative. LSC can be made to fill any wall width that can reasonably dry within the timeframe of most building seasons. Most LSC walls do not exceed a 12″ thickness.

Rounded corners are achievable with LSC
wall systems by using rounded forms.

Straw clay is very fire resistant. Tests conducted by Joshua Thorton and John Straube found that, based on ASTM standards E 111 and E 84, LSC would very likely meet the conditions required for a fire-resistant period of four hours.

They also reported that LSC is a "highly ductile material with the potential to absorb a fair amount of energy in the event of seismic activity."

As each piece of straw has been coated in clay and packed in the wall, there is very little that can actually combust. Although, the walls are breathable to vapor, the continuous wall envelope should not have open channels for sufficient quantities of oxygen to be present, which also helps LSC resist combustion. Like a lot of dense materials, it may only smolder.

According to Franz Volhard, one of the European leaders of earthen construction methods, his own fire tests of LSC demonstrated:

- Light earth responds passively to the Effects of flames, i.e. it does not contribute to the spread of fire.

- The formation of an "insulating" charred layer protects the surface of Underlaying materials from direct exposure to the flame, which increases with flame duration.

- Neither smoke, nor fumes nor perceptible combustion gases were produced.

- No particles fell from the specimens which could have contributed to the spread of the fire.

- Compared with wood-wool magnesite- bonded panel, the fire behavior was better with less charring and no smoke development.

Compressed Earth Block

Earth Blocks, often referred to as compressed earth blocks (CEB) or stabilized compressed earth blocks (SCEB), are a high-quality construction material made from locally sourced soil containing clay. With proper mix design and 4 – 8% cement, Compressed Earth Blocks can meet and exceed compressive strength requirements for cement block in the United States.

Earthen construction has been in practice since the beginning of human civilization. In fact, some of the oldest structures in existence today are made of earth.

Rubble or shallow foundations, which have been in use for centuries, can also be used for CEB building. The advantage of these foundations is that they save money and use less valuable resources (i.e. concrete and fossil fuels) then conventional foundations. For these foundations a trench is dug, filled with gravel/stone and a perforated drain pipe, and then a reinforced grade beam the width of the wall is poured. The perimeter of the foundation trench is usually insulated vertically or horizontally, depending on the design, with rigid foam insulation.

Wall Stacking

CEB walls are stacked in typical masonry fashion with the head joints of the blocks staggered so that the blocks overlap the previous row by at least a third, with the ideal being half. Blocks can be laid either across the width of the wall or turned the other way with the block length running parallel to the wall.

This wall is built with two courses of block running parallel to one another with a space/cavity in between for insulation. Several insulating options are available with this wall system including vegetable based foam, perlite, and sawdust/lime. In this configuration the interior blocks provides thermal mass for heat storage and conservation and the insulated cavity prevent thermal bridging and heat loss to the outside. The exterior course of blocks provides a moderating effect against cold and, even more, against heat gain in the summer.

Whole Wall System

This is simply a solid wall of earth block. It is used for interior walls and for exterior walls when rigid insulation is added to the outside of the blocks.

Hybrid Wall System

One of the unique attributes to CEBs is that they can be incorporated into several hybrid wall systems that are a combination of different building materials and design. For example, CEBs can be used on the interior of a new or existing stick framed home to replace the dry wall and add thermal mass and architectural beauty. CEBs have also been used in conjunction with strawbales, with the bales providing insulation and the CEBs providing thermal mass and the strength and beauty of masonry. In this example the CEBs are placed against the strawbales on the interior.

Mortar

Concrete Bond Beam at top of CEB Wall   CEBs laid in lime mortar

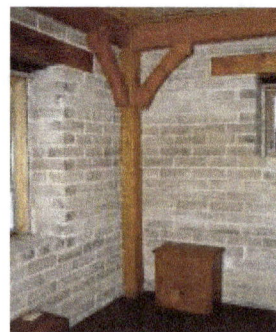

Earth block builders have dry stacked walls, stuck them together with a thin slurry made from the same material as the blocks, and laid them in mortars. We have found that the slurry method works well for unstabilized blocks because it penetrates the surface of the blocks and sticks them together. However, our stabilized blocks are water resistant and stick better with a standard mortar. This mix is made with 2 parts lime to approximately 5 parts 1/8" screened mason's sand or a combination of 1 part lime/1part portland cement to 5 parts sharp sand.

Windows and Doors

Window and door rough openings are built similar to conventional frame construction and placed in the wall according to plans. They can be secured in place with screws through the framing into to the blocks, or expanded metal lathe nailed to the outside of the frame and laid between block courses. Lintels, similar to headers in frame construction, are placed over openings to carry wall and roof loads. Lintels can be made of wood, stone, steel, or concrete.

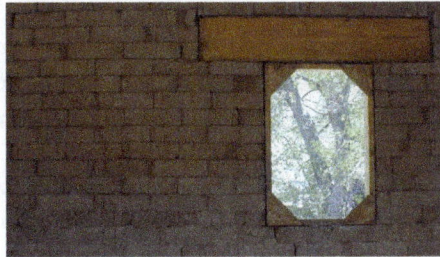

Wood lintel over framed rough opening for window in an un furnished wall

Electrical and Plumbing

Electric wires can be run in the cavity for a double wall system; woven between courses when using a solid wall system; notched into the walls after the walls are built, and then plastered over; or run through conduit in either double wall or solid wall systems. Electric boxes for switches and outlets are placed in the wall and built around.

Plumbing, which is typically run in interior walls, can be plumbed in first and then built around, or run through an interior framed wall or chase.

Bond Beam

A reinforced concrete or wood bond beam is built at the top of a CEB wall to tie all of the walls together, level the walls, and provide an anchor point for the roof or the next story.

Formed Bond Beam

Interior Finishing

Interior walls are typically plastered or left exposed with the blocks visible. Plasters with a base of clay, gypsum, or lime are all appropriate for earth blocks. Earth blocks do not generally take as much time to prep and plaster as other natural building materials. Two coats, a base and then a finish is often enough, although a third coat or color wash is also common. Color is added in the finish coat of plaster, in a wash like an alise (clay paint or lime wash), or breathable natural paints like milk paints.

Once many homeowners see the earth blocks walls, they often ask if they can leave the blocks exposed. No problem, although we do suggest a sealer or wash to eliminate dusting and lighten the color of the blocks.

Dry wall has been put up over earth blocks although this is unnecessary and an added expense. Walls can also be covered with a wainscot, crown molding, or other trim details.

A finished earth block wall can look as smooth as conventional drywall, organic and flowing with ridges and undulations like many naturally built homes, or something in between.

Exterior Finishing

Exterior walls are finished with a stucco or lime plaster/render. Two or three coats of plaster are the typical application systems. There are earth block houses that have been sided with cedar, vinyl, and cultured stone although these are extras and are done so for appearances. Many homeowners like the exposed block look, but we have yet to come across a sealer that stands up to the test of time.

Costs

There are a number of variables that influence the cost of a CEB home such as site location, soil availability and composition, and architectural design. It is important to note that with the ever-increasing cost of fuel and lumber, the CEB cost advantage is continually increasing.

Because CEBs are used entirely as a wall system, the remaining costs, which can represent 80-90% of the total cost of the home, will be the same as conventional building. For example, the cost of the roof, windows, cabinets, etc. are the same for a framed and CEB home.

Building the wall of a home typically represents 10-20% of the total cost of a home. A CEB wall will average 15% more then a conventionally built wall. In other words if the wall of a conventional home cost $15,000 for a $100,000 home, it will cost $2,250 more for a CEB wall. Over the lifetime of the home, this is a small cost when one considers the energy savings, environmental benefits, and aesthetic beauty of an Earth Block home.

Cost on a per block basis average approximately $1.20 per (7" X 14" X 4") stabilized block and $1.00 per unstabilized. A 1000 square foot home will need approximately 5,500 blocks. So, $6,600 or $5,500 would be the block costs of stabilized or unstabilized.

Compressed Earth Block Advantages
- Strong & Durable

Dwell Earth equipment and processes produce the highest quality Earth Blocks in the world. Earth Blocks can meet and exceed concrete block strengths that are produced locally and greatly exceed locally made country fired brick strengths. Our building system is reinforceable allowing modern engineering designs to be applied so that buildings can withstand even the most disastrous circumstances.

- Long History

Earth is the oldest building material known to man. Since the beginning of time people all over the world have been using the dirt under their feet to make quality structures. The oldest standing and continually inhabited buildings in the world are made of earth. Earth Blocks bring the added benefit of engineering and design to modernize this age old building method.

- Healthy

Earth Blocks are all natural and will not emit toxic gases from glues and preservatives as many other building materials do. Earth Blocks create buildings that are able to naturally regulate heat and humidity inside the structures they create. This regulation of humidity greatly increases indoor air quality which is the cause of many common illnesses often attributed to the seasons.

- Sustainability

Earth Blocks have the lowest embodied energy of any building material. They have the smallest carbon footprint and can easily be made to be carbon neutral. Earth Blocks make the most energy efficient structures by taking the strain off of heating and cooling systems.

- Cost Effective

Producing Earth Blocks locally helps to eliminate transportation costs. Cement content is greatly reduced to make Earth Blocks as strong or stronger than other materials available locally. Earth Blocks also reduce lifecycle costs of building operations by creating more energy efficient structures.

- Environmentally Friendly

Compressed earth blocks have the lowest embodied energy of any building material and create structures that are superior in energy efficiency. Earth Blocks are able to regulate temperature and humidity naturally so less energy is required to maintain a comfortable indoor environment.

Sandcrete

Sandcrete block is a composite material made up of cement, sand, water, molded into different sizes. Sandcrete blocks can be made either in solid and hollow rectangular types. They are of sizes and weight that can be easily handled by the bricklayer with the facing surface layer than that of a brick but conveniently dimensioned. The most commonly available sizes are 450 mm x 225 mm x

225 mm and 450 mm x 150 mm x 225 mm. Sandcrete blocks are available for the construction of load bearing and non-load bearing structures.

It is a composition of usually (1:6) mixes of cement and sharp sand with the barest minimum of water mixture, and in some cases admixture, molded and dried naturally. Nigerian Industrial Standard defines sandcrete block as a composite material made up of cement, sand and water, molded into different sizes. Sandcrete blocks constitute a unique class amongst man-made structural component for building in civil engineering work. For example in buildings, walls are constructed using (blocks), as either load bearing or non- load bearing to provide shelter, protection, conveniently divide space, privacy and also to provide security for man and his properties.

Sandcrete is unsuitable for load- bearing columns, and is mainlyused for walls or for foundation if no suitable alternative is available. As material for walls, its strength is less than that of fired clay bricks, but sandcrete is considerably cheaper. Also the time lapse between mixing and compaction has been found to affect the strength. A time lag will not only diminish the hardening effect of the cement but will require extra energy to breakdown the aggregation of particles to achieve the desire density. An increase in strength with age and curing temperature has been reported for cement stabilized sandcrete, but this depends on the nature and texture of sand and the percentage of cement added. Block should be left to mature for at least 28 days (by curing them) before they are laid, if enough strength is needed. Blocks made from a mixture of sand, cement and water are called Sandcrete blocks.

They are used extensively in virtually all African countries including Nigeria. For a long time until perhaps a few years ago these blocks were manufactured in many parts of Nigeria without any reference to any specifications either to suit local building requirements or for good quality work. The high and increasing cost of cement has contributed to the non-realization of adequate housing for both urban and rural dwellers. Alternatives to cement as a material for construction are very desirable in both short and long term as a stimulant for socio-economic development. In the short run, any material that can complement cement and is much cheaper will be of great interest. Over the past decade, the presence of mineral admixtures in construction materials has been observed to impart significant improvement on their strength, durability and workability.

Process of Manufacturing of Sandrete Brick

Sandcrete Blocks: shall mean a composite material made up of cement, sharp sand and water.

i. Blocks shall be molded for sandcrete using metal (wood) molds of:

- 450 mm x 225 mm x 150 mm
- 450 mm x 225 mm x 225 mm
- 450 mm x 225 mm x 100 mm

ii. They are usually joined by mortar, which is a rich mix of sandcrete.

Aggregate

These include both coarse and fine, from natural sources, blast furnace slag, crushed clay and furnace clinker.

Sand

Shall be of approved clean, sharp, fresh water or pit sand, free from clay, loam, dirt, organic or saline water of any description and shall mainly pass 4.70mm test sieve. If lagoon sand is used this must be properly washed to the approval of the supervisor.

Mix Proportion

Mix used for blocks shall not be richer than 1 part by volume of cement to 6 parts of fine aggregate (sand) except that the proportion of cement to mix-aggregate may be reduced to 1:4 ½ (Where the thickness of the web of the block is one 25 mm or less).

Strength Requirements

Sandcrete blocks shall possess resistance to crushing as stated below and the 28day compressive strength for a load bearing wall of two or three story building shall not be less than average strength of 6 blocks, lowest strength of individual block 2.00 N/mm^2 (300psi), 1.75 N/mm^2.

Molding

The 28 day compressive strength of a sandcrete block for load bearing wall of two or three story buildings shall not be less than the values given above and shall comply with the existing NIS specification for sandcrete blocks.

Compaction

Two methods to be applied depending on the availability of materials (tools) are;

1. By approval (standard) machine compaction.

2. My metal mold (hand) compaction.

Production

The sandcrete block shall be cast using an appropriate machine with cement/sand ratio of 1:6 measured by volume. Where hand mixing is carried out, the materials shall be mixed until an even color and consistency throughout is attained. The measure shall be further mixed and water added through a fire hose in such sufficient quantity as to secure adhesion. It shall then be well rammed into molds and smoothed off with a steel face tool.

Materials Selection of Ingredients of Sandrete Brick

Sand

The sand used was clean, sharp river sand that was free of clay, loam, dirt and any organic or chemical matter. It was sand passing through 4.70 mm zone of British Standard test sieves. The sand had a specific gravity of 2.66 and an average moisture content of 0.90%. The coefficient of uniformity of the sand was 2.95.

Cement

The cement used was Ordinary Portland Cement from the West African Portland Cement Company, Ewekoro in Ogun state of Nigeria with properties conforming to BIS 12-1971.

Water

The Water used was fresh, colourless, odourless and tasteless potable water that was free from organic matter of any type.

Sawdust

It was composed of fine particles of wood. The physical and chemical properties of sawdust vary significantly depending on several factors, especially the species of wood. SD used was the mixture of wastes from both hard and soft woods. Preliminary analysis was conducted on the sawdust to determine their suitability for block making. Tests conducted include: particle size analysis of sand, specific gravity test on sawdust and sand. Majority of the fine particles of sawdust passed through 4.76 mm BS test sieve.

Drying and Curing of Sandrete Brick

After removal from machine, the blocks shall be left on pallets under cover in separate rolls, one block high, with a space between each block for at least 24 hours and kept wet by weathering through a fire watering hose. The blocks may then be removed from the pallets and the blocks may be stacked during which time the blocks shall be kept wet. The blocks may be stacked not more than 5 blocks high under cover at least seven days before use after the previous period.

Testing of Sandrete Brick Water Absorption

Water Absorption

Weights should been taken in the dry state and noted, was then fully immersed in water. The time taken for full immersion was noted, and period of twenty-four hours was allowed to elapse. After the 24hours, the wet block samples were the removed and weighed. The difference between the dry and wet weights of each block was the calculated by subtracting the dry weight from the wet weight. From this the water absorption capacity can then be expressed as a percentage i.e.

Wet Weight (Ww) – dry Weight (WD) × 100% / Volume of Block

Bulk Density Determination

The bricks should be labeled and numbered, and they were each weighed in their dry states, during which their masses was read and recorded. The mass scale used was of 50 kg capacity and has 500 g graduations. The dimensions i.e. the length, breadth and height of each block were then taken from this, the volume, and thereafter, the bulk densities were calculated.

Compressive Strength

Three numbers of whole bricks from sample collected should be taken .the dimensions should be measured to the nearest 1mm. Remove unevenness observed the bed faces to provide two smooth parallel faces by grinding. Immerse in water at room temperature for 24 hours. Remove the specimen and drain out any surplus moisture at room temperature. Fill the frog and all voids in the bed faces flush with cement mortar (1cement, 1 clean coarse sand of grade 3 mm and down). Store it under the damp jute bags for 24 hours filled by immersion in clean water for 3 days. Remove and wipe out any traces of moisture. Place the specimen with flat face s horizontal and mortar filled face facing upwards between plates of the testing machine.

Advantage of Sandrete Brick

1. It is cheaper in cost

2. It is much easier to manufacture

3. High compressive strength

4. Its technology is quite cheap

5. Its colour can be easily changed as desired

Wattle and Daub

Wattle and daub walls and the undulations of a distorted roofline form part of the attraction of a medieval timber framed building. The walls gain their character from the timber frame which forms the load bearing structure of the building, leaving open areas between that have to be in-filled to keep the weather out. The type of infilling varies according to the function and status of the building, its location within the country and the locally available materials. It is probably fair to assume that if a material was readily available and could be adapted for use it would have been used as an infill to a timber framed building at sometime, somewhere.

A timber framed building with wattle and daub infill, lime washed in the medieval manner.

Wattle and daub is one of the most common infills, easily recognisable by the appearance of irregular and often bulging panels that are normally plastered and painted. It is an arrangement of small timbers (wattle) that form a matrix to support a mud-based daub. The timbers normally fall into two groups, the primary timbers or staves, which are held fast within the frame and the secondary timbers or withies which are nailed or tied to, or woven around the staves. Arrangement and sizes of panels vary from area to area as does the orientation of the staves. The daub was applied simultaneously from both sides in 'cats' (damp, workable balls) pressed into and around the wattle in order to form a homogeneous mass. As the daub dried it was often keyed by scratching or 'pecking'. Once the daub had hardened, the surface was dampened to receive a lime plaster covering. The surface plaster was usually made of lime and sand or other aggregates reinforced with animal hair or plant fibre. The plaster was finished flush, or in some cases, it would continue across the panels and timbers alike. This would allow less important timbers to be concealed and only principal members to be shown. The plaster may be smooth trowelled, rough cast or even parged (incised and/or built up with a pattern or design).

Performance

The durability of wattle and daub is illustrated by this
wall, still standing after fire burnt the roof off.

Wattle and daub may not be the most rigid material, but therein lies its strength. It is able to accommodate even the most severe structural movement; it is usually well sprung into the timber frame and offers support to weakening timbers that other forms of infill might not. Wattle and daub is not lightweight or flimsy. Its weight is not dissimilar to bricks, however its insulation is better and from a security point of view it can be far more difficult to break through than brick. Although wattle and daub is porous and moisture is absorbed when it rains, moisture levels are kept low because the daub acts like blotting paper to disperse the moisture and because of the high rate of evaporation from its surface.

A wattle and daub panel in need of repair

In moderate, sheltered conditions and if well maintained, a wattle and daub panel should last indefinitely. Examples of 700 years old are known to exist.

Traditional infill panels in timber framed buildings can perform extremely well if properly constructed and maintained. Although in some areas of the country it was normal for infill panels to have protective plaster coatings which extended over the timber frame, it has become fashionable to remove plaster to expose timbers. This is likely to compromise the performance of the building and accelerate the decay of the previously protected structure. It is unreasonable to expect to have a timber frame exposed on both sides and not have draughts and/or some water penetration whether the infill panel is traditional or modern.

The eroded surface of a daub panel revealing its hair and straw binder

Where timber framing was not plastered over it was normal practice to limewash it each spring. Although this was partly for hygienic reasons (being slightly caustic, fresh limewash acts as a mild biocide and disinfectant), it had the tremendous benefit of filling minor cracks caused by seasonal movement. Medieval buildings would have looked quite different from the more recent black and white interpretation that we see so often today.

In some cases weather boarding or tile hanging may have been added over the infill panels, particularly on exposed gables, to protect them from the weather. Removing the protective covering can lead to the recurrence of old problems all over again. It would he wise to learn from our forebear's experience and consider alterations only after careful thought and for good reasons, not purely on aesthetic grounds.

Decay is often caused by the introduction of hard cement in new renders and repairs, and by the use of modern impervious paints. This is because cement based renders are brittle and often crack, especially at the junction with the timber frame. When it rains, water runs down the face of the panels because both the cement and the modern paints are impervious, soaking right into the wall behind wherever a crack is found. Thus the daub will get wetter and wetter over time, leading to the decay of the timber frame and wattles as well as soggy, unstable daub. Only soft, porous and flexible finishes such as haired lime plaster and lime wash should ever be applied to daub.

Repair Considerations

Through the passage of time buildings may become neglected and some damage is inevitable. Knowing whether a damaged panel should be repaired or replaced, even with experience, requires careful consideration, weighing up many factors such as age, importance, rarity, position and function within the building, condition and cost.

Although cost has deliberately been put at the end of this list it will, in many cases, be the deciding factor. Age, importance and rarity can be difficult to define without research, however, bear in mind that all elements of ancient fabric are important and that the loss of any eats away at our heritage.

Repair to Wattles

Repair to daub can normally be implemented, even in the most extreme cases, providing the wattles are still in good condition or repairable. Whereas repair to a panel where the wattles have been totally consumed by fungal decay or insect attack can be very difficult even where much of the daub/plaster survives. Deterioration may be found in the wattles if they have been damp, particularly if they are not oak or contain sapwood, and hazel seems to be particularly prone to decay by woodworm (common furniture beetle). Wattle panels with insect attack may need some localised treatment, but are often strong enough to carry the daub. Introducing additional support can increase their strength. This can take the form of new staves or withies or timber battens or stainless steel mesh fixed across weakened areas. Each repair will be different, depending on the circumstances. In general finding the right solution is a matter of ingenuity based on the defects and conditions found.

Repair to a wattle panel may not be too difficult if the daub has already fallen away. The wattle behind does not need to be absolutely rigid, but should be strong enough to carry the new daub. It may be necessary to hold the wattle firm whilst applying the new daub.

Where daub is still in place the repair of a wattle panel can be much more challenging. In some cases it may be possible to re-support or re-fix loose daub by using non-ferrous wire ties or screws and washers. In some cases it will be necessary to hold a panel carefully in position, or even totally remove it in one piece, while repairs are carried out to the timber frame, and then put it back. In this case specialist advice is essential if a disaster is to be avoided.

Repair to Daub

Some shrinkage is normal even in the most successful of historic daubs, and gaps around the edge of the panel are usually caused by a combination of shrinkage within the daub and the timber frame seasoning. These gaps allow the panel to move, so to keep it weather tight they should be filled. They can easily be filled with daub or lime mortar. If problems are experienced with excessive shrinkage it is either because there is too much suction in the existing daub, or the repair mix is unsuitable, but it is always easier to control the shrinkage of a whole panel with the same moisture content. When areas of daub have failed or become detached, they can be repaired by applying new daub to fill the missing areas (after careful preparation and pre-wetting).

Problems can sometimes be overcome by additional wetting of the existing daub or by modifying the repair mix. The ingredients used in an original daub mix were normally used because they were locally available and cheap, they may not have been ideal. Nevertheless the first recommendation for a compatible material would always be to use the original material. Old daub salvaged from damaged panels can be broken up and mixed with a little water to make it useable again. It may be necessary to add additional material to bulk it out, or modify its performance. However, the required performance of a repair mix may be different from the requirement for a whole panel. A useful tip is to mix one part daub with one part of a good coarse lime mortar to achieve a better-behaved material.

New Daub Mixes

Daub is generally made up of a combination of ingredients shown in the table below.

Binders	Aggregates	Reinforcement	Others
Clay	Earth	Straw	Dung
Lime	Sand	Hair	Blood
Chalk Dust	Crushed chalk	Flax	Urine
Limestone dust	Crushed stone	Hay or grass	Dung

The binder holds the mix together, the aggregates give it bulk and dimensional stability, the reinforcement helps hold it all together, control shrinkage and provide long term flexibility. Some locally available materials may contain more than one of the aggregates and other ingredients. For example, subsoil may contain clay, sand and earth. There is some debate over whether dung was deliberately added to daub mixes. It is probably reasonable to assume that the presence of dung in daub mixes was due to using old straw from animal sheds (why use fresh straw when it is valuable for animal bedding?) and using animals to do the hard work of treading the daub.

Historically, daub was a cheap material and lime was relatively expensive, so it is unlikely that lime was included in daub except under special circumstances. It is far more likely that the expensive lime would have been reserved for the plaster and limewash, where it would be necessary.

There are probably as many daub mixes as there are daub buildings. Try experimenting with locally available materials. Remember to only add enough water to make the mix workable, not so much as to cause excessive shrinkage. Another tip is to mix the ingredients (without hair or straw) in advance and leave the mix to 'temper'. It can then be re-mixed when required and the reinforcement added. This will allow any dry ingredients to soak up water and for the whole mix to have even moisture content.

Repair to Surface Plaster

Where the surface plaster has failed but the daub behind is still sound it is normally possible to repair the plaster. It may be that the whole topcoat to the panel has failed or been removed in the past in which case it will be necessary to replace the whole area. Detached plaster can sometimes be re-secured to the daub behind by means of small stainless steel screws and washers, or re-adhered to the daub surface with a lime mix.

If you are faced with having to repair or replace areas of lime plaster and carry out minor repairs to the daub behind, it may be sensible to consider using lime plaster for the daub repairs as well as the plastering. This is often a sensible approach since it means only having to deal with one type of material and can minimise the shrinkage problems that may occur with small daub repairs.

New woven hazel ready for daubing

Replacement Panels

If a wattle and daub panel is beyond repair or missing altogether then a replacement panel will be required. Before removing any panels of a listed building consult your local conservation officer. Listed building consent will normally be necessary and you may be required to carry out recording of the existing panels before proceeding. Some buildings constructed before the 18th century were decorated with wall paintings to brighten the home. These important works of early art vary from simple patterns of repeated motifs to fine works of art and trompe lloeil architectural elements. Whenever considering the removal of a panel it is essential to be aware that original wall paintings or patterns could be hidden beneath the layers of limewash, plaster or panelling.

Before deciding upon a design for your new panels it is necessary to understand why the old ones have failed and to address these reasons. For example, there is clearly no point in replacing a panel damaged by a leaking gutter if the leak is still there. Wattle and daub is the natural choice for a replacement panel. The evidence of the previous panel will normally dictate the species and pattern of the wattles. Most properly constructed and maintained wattle and daub panels will out-live their builder.

Combining as it does our understanding of traditional performance and the needs of old buildings, wattle and daub has proved itself over time. Properly maintained, the infill panels not only keep the weather out but also create an environment where the structural timber frame is not at risk. Not only is wattle and daub the sound choice from a constructional viewpoint it is also the most environmentally friendly approach. The materials are renewable, from sustainable resources, and minimal energy is consumed in their production.

Construction Aggregate

Aggregates are inert granular materials such as sand, gravel, or crushed stone that, along with water and portland cement, are an essential ingredient in concrete.

For a good concrete mix, aggregates need to be clean, hard, strong particles free of absorbed chemicals or coatings of clay and other fine materials that could cause the deterioration of concrete. Aggregates, which account for 60 to 75 percent of the total volume of concrete, are divided into two distinct categories - fine and coarse. Fine aggregates generally consist of natural sand or crushed stone with most particles passing through a 3/8-inch sieve. Coarse aggregates are any particles greater than 0.19 inch, but generally range between 3/8 and 1.5 inches in diameter. Gravels constitute the majority of coarse aggregate used in concrete with crushed stone making up most of the remainder.

Natural gravel and sand are usually dug or dredged from a pit, river, lake, or seabed. Crushed aggregate is produced by crushing quarry rock, boulders, cobbles, or large-size gravel. Recycled concrete is a viable source of aggregate and has been satisfactorily used in granular subbases, soil-cement, and in new concrete.

After harvesting, aggregate is processed: crushed, screened, and washed to obtain proper cleanliness and gradation. If necessary, a benefaction process such as jigging or heavy media separation can be used to upgrade the quality. Once processed, the aggregates are handled and stored to minimize segregation and degradation and prevent contamination.

Aggregates strongly influence concrete's freshly mixed and hardened properties, mixture proportions, and economy. Consequently, selection of aggregates is an important process. Although some variation in aggregate properties is expected, characteristics that are considered include:

- Grading
- Durability

- Particle shape and surface texture
- Abrasion and skid resistance
- Unit weights and voids
- Absorption and surface moisture

Grading refers to the determination of the particle-size distribution for aggregate. Grading limits and maximum aggregate size are specified because these properties affect the amount of aggregate used as well as cement and water requirements, workability, pumpability, and durability of concrete. In general, if the water-cement ratio is chosen correctly, a wide range in grading can be used without a major effect on strength. When gap-graded aggregate are specified, certain particle sizes of aggregate are omitted from the size continuum. Gap-graded aggregate are used to obtain uniform textures in exposed aggregate concrete. Close control of mix proportions is necessary to avoid segregation.

Common Aggregates

- Crushed Stone and Manufactured Sand

Stone is quarried, crushed and ground to produce a variety of sizes of aggregate to fit both 'coarse' and 'fine' specifications.

- Gravel

Gravel is formed of rocks that are unconnected to each other. 'Gravel is composed of unconsolidated rock fragments that have a general particle size range and include size classes from granule- to boulder-sized fragments.'

- Sand

Sand occurs naturally and is composed of fine rock material and mineral particles. Its composition is variable depending on the source. It is defined by size, being finer than gravel and coarser than silt.

- Lightweight Aggregates

Vermiculite

Glass aggregate

Lightweight aggregates can be from natural resources, or they can be man-made. The major natural resource is volcanic material whilst synthetic aggregates are produced by a thermal the thermal treatment of materials with expansive properties.

These materials can be divided in three groups—natural materials, such as perlite, vermiculite, clay, shale, and slate; industrial products, such as glass; and industrial by-products, such as fly ash, expanded slag cinder, and bed ash.

- Recycled Concrete

Recycled concrete is created by breaking, removing, and crushing existing concrete to a preferred size. It is commonly used as a base layer for other construction materials.

Recycled concrete can be used as aggregate in new concrete, particularly the coarse portion. When using the recycled concrete as aggregate, the following should be taken into consideration:

- Recycled concrete as aggregate will typically have higher absorption and lower specific gravity than natural aggregate and will produce concrete with slightly higher drying shrinkage and creep. These differences become greater with increasing amounts of recycled fine aggregates.

- The chloride content of recycled aggregates is of concern if the material will be used in reinforced concrete. The alkali content and type of aggregate in the system is probably unknown, and therefore if mixed with unsuitable materials, a risk of alkali-silica reaction is possible.

Aggregate Extraction

Aggregates are extracted from natural sand or sand-and-gravel pits, hard-rock quarries, dredging submerged deposits, or mining underground sediments.

Rock Quarries

The process of extraction from rock quarries usually involves explosives to shift the rock from the working face. Rock is crushed and passed through a series of screens. The output is a range of sizes of rock produced to specified sizes. Crushed rock is transported from quarries by road or rail.

Sand and Gravel Quarries / 'Pits'

Reflecting the essential nature of the material, sand and gravel quarries, both working and defunct, are a common feature of the UK landscape, particularly in the East of England. Pits are located in areas where glaciers left behind clean deposits of sand and stone. Sometimes the gravel is deeper than the groundwater table and the gravel is extracted through pumping - leaving behind ponds and lakes.

Marine Aggregate

Much of today's seabed was dry land 20, 000 years ago when sea levels were up to 100m lower. After the last Ice Age sea levels began to rise and existing river valleys bearing sand and gravel deposited by glaciers became submerged. Eventually sea levels rose to establish today's coastline. The former river sediment has been re-worked by the action of the sea to leave clean and well-sorted aggregates.

Between 20 and 30 purpose-built dredging vessels work 24/7 to extract marine aggregate. There are two types of dredging technique:

- Static dredging involves a vessel anchoring over and working a deposit using an electronic pump.
- A pump is trailed behind the vessel along the seabed.

Water

Aggregate washing

Water is critical in the making of concrete. Adding water to the mix sets off a chemical reaction when it comes into contact with the cement. The water used in the mixing of concrete is usually of a potable standard. Using non-drinking water or water of unknown purity risks the quality and workability of the concrete.

Light Expanded Clay Aggregate

Light expanded clay aggregate, also called lightweight expanded clay aggregate, or LECA, is a form of high-temperature burnt clay nodules. LECA is used for a very wide range of purposes, many of them in agricultural and hydroponic systems.

Sometimes referred to as clay pebbles or clay pellets, LECA is considered a soilless growing medium when used alone in hydroponics. It can also be used to amend soil.

Also known as LECA and sometimes called lightweight expanded clay aggregate, light expanded clay aggregate takes the form of small balls or pellets. These are formed from special "plastic" clay that is fired in a rotary kiln. During the kilning processes, gases created released by the heat warm and expand, inflating the balls and forming a honeycomb structure.

Light expanded clay aggregate is strong, durable, and is an excellent solution to many challenges faced by gardeners and indoor growers alike. LECA can be used in heavy soils to prevent compaction and enhance aeration, for instance. They are also be used in hydroponics as a stand-alone growing medium.

When mixed with soil and/or peat, LECA helps enhance drainage. However, they also absorb and retain water, meaning that they can help ensure healthy plant growth during dry periods of the year. These clay balls also retain heat very well, which makes them an excellent option for insulating plant roots during cooler periods of the year.

Other qualities that make LECA a popular option with growers include the fact that this grow medium is completely natural and is not subject to dry-rot or wet-rot, the way many other substrates are. It is also inflammable.

Properties

Lightness:

Lightness of Leca is for the multi-separated air spaces which exist inside and among the aggregates, density of aggregates, depending on the size of Leca, Is fluctuated from 380 to 710 kg/m³.

Thermal Insulation:

Thermal conductivity coefficient for LECA aggregate is about $0.09 < \lambda < 0.101$ and for Leca concrete 0.208 w/mc° with density of 800 kg/m³. the results have tested for the high quality for insulating feature of Leca products.

Sound Insulation:

Leca aggregate and block are among the best insulating materials and according to conducted tests, sound in Leca blocks with 10 cm width and 15 cm width respectively fades 45db and 46 db and refers to relavant standards codes are suitable for most of purposes.

Fire Resistance:

Since Leca aggregates are exposed to 1200 degree centigrade temperature in rotary kiln, these aggregates are suitable fire resistance and have already passed the practical test during the producing process. Walls made of Leca blocks (about 130 kg/m²) can resist 3 hours against fire and fire stretch which should be considered a crucial safety element.

Non-decomposability:

Leca aggregate strongly resists against alkaline and acidic substances and pH of nearly 7 makes it neutral in chemical post reaction with concrete.

Water Absorption:

For closed inside porousness, average water absorption of Leca aggregate (0-25 mm) is about 18 percent of volume in saturated state during 72 hours.

Leca wide applicability	Leca Gradation	
Leca Light Weight Concrete, light weight block, Prefabricated Panels&Slabs. Light Filler, Leca Mortar and Water Purification system. Agriculture & Aquaculture.	0-4 mm	< 710 kg/m^3
Light Weight Concrete, light weight block, Prefabricated Panels & Aquaculture, Ornamentation.	4-10 mm	< 480 kg/m^3
Lightweight Filler Concrete, Sewage system. Landscaping, Agriculture and Aquaculture, Drainage.	10-25 mm	< 380 kg/m^3
Floor and Roof sloping, Light weight Filler, Road Construction.	0-25 mm	< 430 kg/m^3

Leca in Construction

Lightweight Aggregates Concrete

In all the cases which lightening and insulation should be considered in buildings, Leca light weight aggregate (structural& non- structural) comparing to autoclaved aerated concrete foam concrete or concrete made of pumice or scoria has a wider range of applicability.

Leca Lightweight Concrete Block

Leca block (back) is produced by mixing of Leca aggregates, Cement, sand and water. Using of Leca aggregate decreases the concrete density.

Advantages:

1. Lightening up to 30% of dead load.

2. Self thermal & Sound Insulation.

3. Appropriate Behavior in Earthquake:

Low Module of elasticity of Leca Concrete, and having cement base material as well as its mortar, in addition to the lightness, reduces the destructive effects of earthquake on structure. Shapes of Leca blocks are designed for conducting vertical mortar in joints which minimizes the rubble threat. Referring to mentioned reasons, using of Leca blocks prevents buildings from asymmetric settlement and other detrimental side effects.

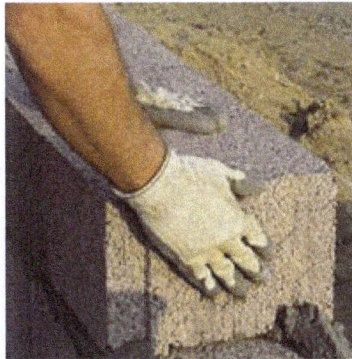

4. Physical Operation:

Leca Blocks are suitable for any physical operation like cutting, nailing, transfixing and ridge extending to make proper route for wire, pipe and othe installing components without any cracking.

5. Accelerating Construction, Saving time, Labor, and material:

For lightness and different sizes and dimensions of Leca blocks, walling up could be done in high speed which leads to save time, labor and material.

6. Safe for Installation:

Being chemically neutral, Leca aggregate and blocks are the best bed for building installation and prevent pipe and wires from decay.

Leca blocks are designed in a way not to be only lightweight but cover all required.

Flooring and Roofing

Lightness, insulating, durability, non-decomposability, structural stability, and chemical neutrality features are collected in Leca as the best lightweight aggregate for flooring and roofing and in the most suitable and reasonable replacement for pumice, scoria and polystyrene, moreover Leca aggregates are used to mark insulating lightweight concrete for all floors substructures under stone, marble or ceramic etc.

Leca built home keeps you warm in winter and cold in summer.

Gypsum Block

The gypsum blocks are small building items – 3 per m² – designed for the building of partition walls. The tongues and grooves moulded on the edges make the assembly easy, quick and robust. The perfect flatness of the gypsum blocks also enables to avoid using coatings.

Gypsum block is natural gypsum or chemical gypsum used as the main raw material, which is mixed with water, casted and dry bulk. It is made of lightweight wall materials and its shape is rectangular, vertical and horizontal edges, respectively, with tenon and mortise. It was added to allow the production of fiber-reinforced material or other aggregates, blowing agents may also be added, water repellent. For industrial and civil buildings in non-load-bearing walls, with a simple structure, the wall is thin, good water resistance, and low production cost.

Foreign gypsum block originated in the mid-20th century, tens of millions of the world's production meters, including France, the annual output of 17 million square meters, while the domestic late start, and the production-based hand, the current domestic production of about 3 million square meters, while China's building habits and structures are very similar and Europe, as a new generation of wall materials, gypsum block in the country will certainly be a very strong market prospects.

Here, the performance characteristics of gypsum block and part of the analysis and comparison of the existing main wall material, the following data according to the Beijing Municipal Commission of Housing and Urban "new technology" and "gypsum block Application Manual" in Data provided by the order:

1. Weight

As a lightweight wall materials, weight is one of its most important features, gypsum block hollow plaster can be divided into two blocks and solid blocks of gypsum, according to JCT698-2010 "gypsum block" standard, currently hollow block table apparent density of 400 ~ 700kg / m³, solid block of apparent density of not more than 1000kg / m³, less than the weight of a single block of 30kg, easy to carry an adult, in the construction without the help of other construction machinery handling, improve the speed of masonry. Quality gypsum block masonry wall unit area, but clay hollow blocks, GRC light panels, fly ash aerated concrete block, aerated concrete about 1/2, so that the whole building loads significantly reduced.

2. Construction operability and ensure the quality of the wall

Gypsum block having sawing, nailing, planing and drilling, bonding and other workability characteristics, convenient and quick dry construction, no construction waste, can be repaired without reinforcement keel, do not wipe the surface decoration, construction High efficiency. A worker can be paved daily 20-40 m² gypsum block walls, because without the required lifting equipment, partition less hardware and spare parts, simple operation. And the wall or other pipelines buried wire is relatively simple, to use the tool in the back wall openings or slotted An End pipeline or socket devices can be closed with putty. Completely overcome the past due to limit the impact of the wall structure, resulting in detours wire loop, in the wall around the drill, drill holes and other defects, indoor electric line laying, installation provided the conditions.

Aerated concrete walls according to national quotas per cubic artificial as 1.475, according to 100mm thick and converted into man-days per 6.8 square masonry, masonry rate far below the average 20 square meters of gypsum block.

3. Fire performance

Gypsum block because of its special molecular structure ($CaSO_4 \cdot 2H_2O$) containing water of crystallization, the temperature rise can release water to 80mm block, for example, gypsum block per m2 will be released upon combustion steam 15kg, one side 3m length of 5m high wall can release after 200kg water temperature will continue to rise, anhydrous calcium sulfate formed after dehydration is a good thermal insulator nonflammable, can effectively prevent the spread of fire, with special fire performance.

4. Healthfulness

With the improvement of people's material and cultural life, health is one of the most frequent word appears in today's society, food and health-related living lines, in the past our building health requirements are very low, or that there is no requirement, but with the community development, construction and health is increasingly becoming issues of concern, because of its non-toxic plaster, the advantages of air humidity regulator in the world's most environmentally friendly construction materials have to say, and aerated concrete in one of the main components of fly ash containing mercury, arsenic, lead, cadmium and other toxic elements and immeasurable natural radioactive elements, also contains high active indirect carcinogen benzopyrene.

5. Sound insulation performance

120 mm thickness of the wall in the same terms, gypsum block (without plaster) weighted sound reduction Rw is 45.5 dB, double-sided plaster (10 mm) of fly ash aerated concrete block is 40 dB, 240 mm double-sided plaster (10 mm) of fly ash aerated concrete block is 47 dB, while 200 mm gypsum block wall is 52 dB, can be seen, in terms of sound insulation, gypsum block for aerated concrete block has a great advantage.

In addition, the weighted sound 100mm and 150 mm double-sided plaster gypsum block (10 mm) of fly ash aerated concrete blocks are 44 dB, which means that when you do within the walls, 100mm gypsum block wall fly ash can replace 170 mm block walls, interior walls can save 0.07 m² per meter indoor area, a 100 m² storey residential 3 m can save area 5 m².

6. Thermal insulation properties

When a certain thickness of the wall, its smaller thermal conductivity, thermal insulation coefficient greater. After hardening plaster because of a microporous structure, with good thermal insulation can, its thermal conductivity is 0.17 Kcal / m² · h ·°C is only 1/3 of brick, one-fifth ordinary concrete, stone 1 /8.

7. Wall Stability

Gypsum block wall material is a ductile characteristics, volume stability in the frame structure of gypsum maintain long-term close connection, gypsum block wall can be synchronized with the deformation of the frame structure. Because the characteristics of volume stability, the connection as long as gypsum blocks and frame structure between the measures in place, we can effectively prevent dislocation between the wall and the frame seismic.

Gypsum block its construction materials and masonry materials are used in gypsum-based products, non-hot expansion difference, and other materials, such as fly ash aerated block, due to thermal performance values between the substrate and the mortar varies greatly , drying shrinkage stress is large (hot red Relay), aerated concrete block and brick and mortar due to temperature is between 0.3mm / m ~ 0.5mm / m, is four times as common clay brick, linear expansion coefficient much larger than ordinary bricks, resulting in a large amount of deformation, resulting in wall cracks.

Capillary suction effect aerated concrete block is poor, single-ended to slow down after water, water for a long time, 24h within the water quickly, after gradually slow, know 10D to reach equilibrium, but not much volume, so if the grass handled properly, will continue to absorb moisture in the mortar, mortar does not meet the conditions before the intensity will lose hydration, resulting in hollowing cracking plaster.

8. Environmental protection

Gypsum products are affected generating mechanism, is a recyclable material, namely plaster products into the thickness of the wall masonry, if pre-demolition, gypsum block can be removed to the gypsum flour mill wastes for recycling, according to the laboratory As a result, so back and forth 10 times, plaster strength decreased only 4%, while including fly ash aerated block, including other wall materials are not reversible reaction, can only be formed after the demolition of the wall construction waste.

9. Cost

Gypsum block with 100 mm and 150 mm aerated fly ash blocks in Tai'an market as an example for analysis, the two materials are currently Tai'an ex-factory price 21 yuan / m² and 20.3 yuan / m² (135 元 / m³), gypsum puzzle block into a wall Price (artificial materials) 30 yuan / m², fly ash brick masonry costs (including artificial materials) 18 yuan / m², single plastering costs (including artificial materials) 30 yuan / m², so that each square meter reached levels can be linked into a wall putty, gypsum block wall into a total cost of 51 yuan / m², fly ash blocks into the wall total cost of 98.3 yuan / m². If all calculated to 150 mm thickness, gypsum block material price is 28 yuan / m², masonry costs 33 yuan / m², still save 37.3 yuan / m² ratio of aerated concrete, visible, gypsum block unmatched in the promotion of market price advantage.

References

- Types-of-building-materials-construction-699: theconstructor.org, Retrieved 09 March 2018
- Cob-houses: cobcourses.com, Retrieved 24 May 2018
- The-advantages-and-appropriate-use-of-light-straw-clay, building science: greenbuildermedia.com, Retrieved 26 May 2018
- Wattle-daub-home: earthhomesnow.com, Retrieved 10 April 2018
- Aggregates-for-concrete, building-design: greenspec.co.uk, Retrieved 20 July 2018
- Light-expanded-clay-aggregate-leca: maximumyield.com, Retrieved 22 April 2018

2

Concrete and Cement

Concrete is a construction material that is composed of cement, sand and coarse aggregates that when mixed with water, hardens with time. A commonly used form of cement is the Portland cement. It is used for building beams, slabs, columns and foundations. This chapter closely examines the different types of cement and concrete used in construction, such as Gypsum concrete, precast concrete, Roman concrete, Portland cement, Geopolymer cement, fibre cement, etc.

Concrete

Concrete is a construction material composed of cement, fine aggregates (sand) and coarse aggregates mixed with water which hardens with time. Portland cement is the commonly used type of cement for production of concrete. Concrete technology deals with study of properties of concrete and its practical applications.

In a building construction, concrete is used for the construction of foundations, columns, beams, slabs and other load bearing elements.

There are different types of binding material is used other than cement such as lime for lime concrete and bitumen for asphalt concrete which is used for road construction.

Various types of cements are used for concrete works which have different properties and applications. Some of the types of cement are Portland Pozzolana Cement (PPC), rapid hardening cement, Sulphate resistant cement etc.

Materials are mixed in specific proportions to obtain the required strength. Strength of mix is specified as M5, M10, M15, M20, M25, M30 etc, where M signifies Mix and 5, 10, 15 etc. as their strength in kN/m^2. In United States, concrete strength is specified in PSI, which is Pounds per Square Inch.

Water cement ratio plays an important role, which influences various properties such as workability, strength and durability. Adequate water cement ratio is required for production of workable concrete.

When water is mixed with materials, cement reacts with water and hydration reaction starts. This reaction helps ingredients to form a hard matrix that binds the materials together into a durable stone-like material.

Concrete can be casted in any shape. Since it is a plastic material in fresh state, various shapes and sizes of forms or formworks are used to provide different shapes such as rectangular, circular etc.

Various structural members such as beams, slabs, footings, columns, lintels etc. are constructed with concrete.

ACI 318 Building code requirements for structural concrete and ACI 301 Specifications for Structural Concrete are used in United States as standard code of practice for concrete construction.

There are different types of admixtures which are used to provide certain properties. Admixtures or additives such as pozzolans or superplasticizers are included in the mixture to improve the physical properties of the wet mix or the finished material.

Various types of concrete are manufactured these days for construction of buildings and structures. These have special properties and features which improve quality of construction as per requirement.

Components of Concrete

Components of concrete are cement, sand, aggregates and water. Mixture of Portland cement and water is called as paste. So, concrete can be called as a mixture of paste, sand and aggregates. Sometimes rocks are used instead of aggregates.

The cement paste coats the surface of the fine and coarse aggregates when mixed thoroughly and binds them. Soon after mixing the components, hydration reaction starts which provides strength and a rock solid concrete is obtained.

Grade of Concrete

Grade of concrete denotes its strength required for construction. For example, M30 grade signifies that compressive strength required for construction is 30MPa. The first letter in grade "M" is the mix and 30 is the required strength in MPa.

Based on various lab tests, grade of concrete is presented in Mix Proportions. For example, for M30 grade, the mix proportion can be 1:1:2, where 1 is the ratio of cement, 1 is the ratio of sand and 2 is the ratio of coarse aggregate based on volume or weight of materials.

The strength is measured with concrete cube or cylinders by civil engineers at construction site. Cube or cylinders are made during casting of structural member and after hardening it is cured for 28 days. Then compressive strength test is conducted to find the strength.

Regular grades of concrete are M15, M20, M25 etc. For plain cement concrete works, generally M15 is used. For reinforced concrete construction minimum M20 grade of concrete are used.

Concrete Grade	Mix Ratio	Compressive Strength MPa (N/mm²)	psi
Normal Grade of Concrete			
M5	1 : 5 : 10	5 MPa	725 psi
M7.5	1 : 4 : 8	7.5 MPa	1087 psi
M10	1 : 3 : 6	10 MPa	1450 psi
M15	1 : 2 : 4	15 MPa	2175 psi
M20	1 : 1.5 : 3	20 MPa	2900 psi
Standard Grade of Concrete			
M25	1 : 1 : 2	25 MPa	3625 psi
M30	Design Mix	30 MPa	4350 psi
M35	Design Mix	35 MPa	5075 psi
M40	Design Mix	40 MPa	5800 psi
M45	Design Mix	45 MPa	6525 psi
High Strength Concrete Grades			
M50	Design Mix	50 MPa	7250 psi
M55	Design Mix	55 MPa	7975 psi
M60	Design Mix	60 MPa	8700 psi
M65	Design Mix	65 MPa	9425 psi
M70	Design Mix	70 MPa	10150 psi

Making Concrete

Concrete is manufactured or mixed in proportions with respect to cement quantity. There are two types of concrete mixes, i.e. nominal mix and design mix. Nominal mix is used for normal construction works such as small residential buildings. Most popular nominal mix is in the proportion of 1:2:4.

Designs mixed concrete are those for which mix proportions are finalized based on various lab tests on cylinder or cube for its compressive strength. This process is also called as mix design. These tests are conducted to find suitable mix based on locally available material to obtain strength required as per structural design. A design mixed offers economy on use of ingredients.

Once suitable mix proportions are known, then its ingredients are mixed in the ratio as selected. Two methods are used for mixing, i.e. Hand Mixing or Machine Mixing.

Based on quantity and quality required, the suitable method of mixing is selected. In the hand mixing, each ingredients are placed on a flat surface and water is added and mixed with hand tools. In machine mixing, different types of machines are used. In this case, the ingredients are added in required quantity to mix and produce fresh concrete.

Once it is mixed adequately it is transported to casting location and poured in formworks. Various types of formworks are available which as selected based on usage.

Poured concrete is allowed to set in formworks for specified time based on type of structural member to gain sufficient strength.

After removal of formwork, curing is done by various methods to make up the moisture loss due to evaporation. Hydration reaction requires moisture which is responsible for setting and strength gain. So, curing is generally continued for minimum 7 days after removal of formwork.

Types of Concrete Construction

Concrete is generally used in two types of construction, i.e. plain concrete construction and rein-forced concrete construction. In PCC, it is poured and casted without use of any reinforcement. This is used when the structural member is subjected only to the compressive forces and not bending.

When a structural member is subjected to bending, reinforcements are required to withstand tension forces structural member as it is very weak in tension compared to compression. Generally, strength of concrete in tension is only 10% of its strength in compression.

It is used as a construction material for almost all types of structures such as residential concrete buildings, industrial structures, dams, roads, tunnels, multi storey buildings, skyscrapers, bridges, sidewalks and superhighways etc.

Example of famous and large structures made with concrete are Hoover Dam, Panama Canal and Roman Pantheon. It is the largest human made building materials used for construction.

Steps of Concrete Construction

The construction steps are:

1. Selecting quantities of materials for selected mix proportion

2. Mixing

3. Checking of workability

4. Transportation

5. Pouring in formwork for casting

6. Vibrating for proper compaction

7. Removal of formwork after suitable time

8. Curing member with suitable methods and required time

Properties of Concrete

For a specific type of structure, certain characteristics of concrete to be used may be more im-portant than others. For example, concrete for a multistory building or bridge should have high compressive strength, whereas concrete for a dam should be more durable and watertight whereas strength can be relatively small.

Workability

It is the ease with which concrete can be placed & finished. It is an important property for many applications of concrete. One characteristic of workability is consistency or fluidity which can be measured using slump test.

In the slump test, a specimen of concrete mix is placed in a mould shaped as the frustum of a cone, 12 in high, with 8-in diameter at base and 4-in diameter at top. When the mould is lifted up the change in height of specimen is measured. This change in height is taken as the slump value. Higher is the water content larger is the stump value.

Durability

Concrete should be capable of withstanding the weathering effects, chemical action and should be able to resist load to which it will be subjected in service life. Much of the weather damage sustained by concrete is attributable through freezing and thawing cycles. Resistance of concrete to such damage can be improved by increasing the water tightness.

Water Tightness

It is an important property of concrete that can be improved by reducing the amount of water in the mix. Excess water leaves voids and cavities after evaporation, and if they are interconnected, water can penetrate or pass through the concrete. Prolonged thorough curing as well as entrained-air (minute bubbles) usually increases water tightness. Water tightness can be increased by improving effective compaction of concrete, controlling aggregate grading, using construction chemicals etc.

Strength

This property is usually of main concern. Normally it is determined by knowing the ultimate strength of a specimen in compression tested in the lab but sometimes flexural or tensile capacity is also important which can also be determined through lab tests. Since concrete usually gains strength over a long period, (90 days) the compressive strength at 28 days is commonly used as a measure of this property. Concrete strength is influenced mainly by the water cement ratio, mix proportions and other factors.

Gypsum Concrete

Gypsum concrete is a type of concrete, which is made with a base of gypsum binding materials (for the most part, structural gypsum). Rock mineral aggregates (primarily those with a porous and rough surface) and organic aggregates (such as wood shavings, sawdust, and chopped straw) are used in making gypsum concrete. Additives are introduced into gypsum concrete that retard its setting and also increase its resistance to water and the atmosphere. The strength of gypsum concrete depends on the same factors that determine the strength of ordinary cement concrete.

Composition

US patent 4,444,925 lists the components of Gyp-Crete as atmospheric calcined gypsum, sand, water, and small amounts of various additives. Additives listed include polyvinyl alcohol, an extender such as sodium citrate or fly ash, a surfactant such as Colloid defoamer 1513 DD made by Colloids, Inc., and a fluidizer based on sodium or potassium derivatives of naphthalene sulfonate formaldehyde condensate. One example mix is shown below.

Component	Amount	Approximate Percentage
Water	6.5-8.5 gal	19%
Sand	150-200 lbs	57%
Polyvinyl Alcohol	0.45 lbs	0.14%
Fluidizer	108.8 gr	0.0047%
Extender	22.27 gr	0.00098%
Atmospheric calcined gypsum	80 lbs	24%

The purpose of the polyvinyl alcohol is to prevent the surface of the concrete from becoming dusty. While the exact mechanism is not known, it is thought that as the concrete sets, water migrates to the surface, bringing with it fine, dusty particles. When the water evaporates, the dusty particles are deposited on the surface. It is thought that the polyvinyl alcohol prevents the dusty particles from migrating upwards with the water.

The mix is prepared on site using a specialized truck. The truck contains a tank for water, a mixing tank, a holding tank, a pump, and a conveyor for the sand and calcined gypsum. A hopper for the sand and gypsum is mounted externally on the vehicle.

To prepare the mix, the sand and calcined gypsum are added to the hopper and mixed. Most of the required water is added to the mixing tank, then the sand and calcined gypsum are mixed in. Once all the sand and calcined gypsum have been mixed in, the rest of the water is added until the proper consistency is attained. Finally, the additives are mixed in and the whole batch of concrete is moved to the holding tank to be pumped out into the required area via long hoses. A small sample is taken from the batch and set aside so that the set-up time can be observed and adjustments can be made to the amount of additives so that the timing is correct.

Once the mix has been poured, little leveling, if any, is needed. The mix should be smoothed gently with a flat board, such as a 40" 1x4. This helps to concentrate the calcined gypsum at the surface.

Previous Formulations

US patent 4,075,374 lists the formulation as 10 parts pressure calcined gypsum, 38-48 parts sand, and 4-10 parts water. 0.03 to 0.1 parts of a latex emulsion, such as Dow Latex 460, were also added. To prevent foaming, a defoamer such as WEX was added to the latex at a concentration of 0.2%. It was stated that gypsum calcined at atmospheric pressure produced poor results due to it having flaky particles, and that gypsum calcined under a pressure of 15-17 psi produced better results because it had denser, crystalline particles.

Later it was found that this original formulation expanded too much and in some instances floors cracked. US patent 4,159,912 describes changes made so that the expansion was greatly reduced. In that formulation, 5-8% of Portland cement was added to reduce the expansion. The latex emulsion and antifoaming agent were no longer necessary as the concrete was strengthened by the Portland cement. It was found that atmospheric calcined gypsum could be used for the majority of the calcined gypsum if it was ball milled to change the texture. The proportion of sand was also changed, so that it was in a 1:1.3 to 1:3 ratio with the calcined gypsum. This resulted in a runnier mix, but the set up time was not changed.

Advantages and Disadvantages

Gypsum concrete is lightweight and fire-resistant. A 1.5-inch slab of gypsum concrete weighs 13 pounds per square foot versus 18 pounds per square foot for regular concrete. Even though gypsum concrete weighs less, it still has the same compressive strength as regular concrete, based on its application as underlayment or top coat flooring. A 7-man work crew can lay 4–6 times as much gypsum concrete in a workday as regular poured Portland cement. This is due to the ease of leveling the very runny gypsum concrete versus normal concrete. In addition, if the wooden subfloor is first coated in a film of latex, the adhesion between the subfloor and the concrete is much better than the adhesion obtained with "normal" concrete. A further benefit is that nails can be driven through the cement into the subfloor without it chipping. The cost of gypsum concrete is comparable to regular concrete, ranging from $1.75 per square foot to $6.00 per square foot. Regular concrete ranges from $2.50 to $4.50 per square foot.

Precast Concrete

Precast concrete is widely used in low- and mid-rise apartment buildings, hotels, motels, and nursing homes. The concrete provides superior fire resistance and sound control for the individual units and reduces fire insurance rates.

Precast concrete is also a popular material for constructing office buildings. The walls of the building can be manufactured while the on-site foundations are being built, providing significant time savings and resulting in early occupancy.

The speed and ease with which precast structures can be built has helped make precast a popular building material for parking structures. Precast concrete allows efficient, economical construction in all weather conditions and provides the long clear spans and open spaces needed in parking structures. For stadiums and arenas, seating units and concrete steps can be mass produced according to specifications, providing fast installation and long lasting service. In addition, pedestrian ramps, concession stands, and dressing room areas can all be framed and constructed with precast concrete.

The smooth surfaces produced with precast concrete and the ability of precast, prestressed concrete to span long distances makes precast suitable for use in manufacturing and storage structures. Additional applications for precast concrete include piles and deck for railroad and highway bridges, railway crossties, burial vaults, educational institutions, commercial buildings such as shopping malls, and public buildings including hospitals, libraries, and airport terminals.

Precast Concrete Manufacturing

Precast concrete is created off-site using a mold. That's the main difference between precast concrete and site cast concrete, which is poured into its final destination on site. Here is a simplified overview of the precast concrete process:

- Precast concrete is poured into a wooden or steel mold with wire mesh or rebar. This mold may also have prestressed cable, if needed.

- It is cured in a controlled environment — usually at a plant.

- Once finished, the precast concrete is transported to a construction site and put into place.

It's important to note that not all precast concrete is prestressed with cable reinforcement. The addition of this reinforcement is particularly useful in many structures and buildings where maximizing the strength of the concrete is essential. The addition of the wire or rebar provides tension within the concrete, which is released once curing is complete. The release of the wire or rebar tension transfers strength to the concrete, creating an even stronger material.

Regardless of whether or not prestressing is a part of the equation, this process is faster, safer and more affordable than standard concrete. Precast concrete materials help we maximize your project's potential while making sure it is completed on time. They are also among the most versatile products in construction, combining a strong structure with the ability to:

- Choose any combination of color, form or texture

- Integrate facades

- Meet compatibility requirements for historic structures

- Create everything from small sections to long open spans

- Be recycled or reused upon removal or replacement

Types of Projects Using Precast Concrete

Perhaps the versatility is one of the reasons precast concrete structures are so diverse — ranging from parking garages, bridges and office buildings to stadiums, retail shops and housing. It's clear any number of building types can benefit from the advantages of precast concrete products. Some of the most common construction projects that use precast concrete are listed below.

1. Precast Concrete Structures

Since durability is one of the key characteristics of concrete construction, it's no surprise that many precast concrete structures are used in applications that see a lot of wear and tear from everything from traffic to weather elements. Going hand-in-hand with durability is its strength — another reason it is especially popular for these applications.

- Parking Structures: In parking structure design, durability, economy and installation are three key points of consideration, which is why precast concrete is usually the building material of choice. You'll find several different precast concrete products in parking garages — columns, traffic barriers, stairs, paving slabs, architectural veneer and more. Precast concrete is useful for single-level parking structures as well as larger and more elaborate mid-rise structures. The Pier Village Parking Structure is an example of a precast concrete parking structure.

- Foundations: Precast concrete is used to create entire buildings but in cases where it isn't utilized for the entire building, it may still be used for the foundation. Many residential homes and other buildings have precast concrete foundations, regardless of what is used for walls and floors in the rest of the building. Its reputation for providing a moisture-free, and energy-efficient basement is often what makes precast concrete the material of choice.

- Bridges: The Walnut Lane Memorial Bridge began the precast concrete industry in the United States, and using precast concrete materials for bridges continues today. You'll find precast concrete materials are used for beams, arches, girders, deck slabs, caps and more. Regardless of the size of the bridge, precast concrete gives engineers the ability to create a structure that blends in with the environment and is compatible with any historical surroundings.

- Culverts: When we remember the underground tunnels of ancient Rome are suspected to be early signs of precast concrete, it's easy to see how a section of modern-day underground infrastructure is the perfect application for precast concrete. Box and three-sided culverts are manufactured in all different shapes and sizes to aid in stormwater and wastewater drainage, create short bridges, and retain rainwater and more. Many of them are built using precast concrete to ensure a high-quality and durable product that can be installed efficiently.

- Curb Inlets and Catch Basins: Just like culverts are a part of the underground infrastructure, so are curb inlets and catch basins for wastewater management. Different states and local municipalities have different standards for these pieces, but precast concrete manufacturing can take all of them into consideration and create a product that helps stormwater runoff drain to the underground infrastructure in place.

- Sound Walls: In urban areas, sound walls are erected as a noise barrier between highways and communities. Using precast concrete for these structures can cut noise pollution up to 50 percent. The versatility of design enables these sound wall structures to blend into their surroundings with a specific color, texture or design.

- Retaining Walls: Many precast concrete retaining systems include segmental retaining wall (SRW) products, large precast modular blocks (PMB), mechanically stabilized earth (MSE) face panels, crib walls, cantilever walls and post-and-panel systems. Each of these elements specifications easily met in a timely fashion by precast concrete.

2. Precast Concrete Buildings

The fire-resistant and sound-attenuating characteristics of precast concrete products make them ideal for a variety of building applications. Reducing moisture and creating an energy efficient environment are two other convincing factors when considering a precast concrete building. The diverse variety of buildings included below encompasses the versatility of precast concrete, as these materials come together to create an impressive result.

- Office Buildings: The unique characteristics of precast concrete products allow for unique building designs that are attractive and functional. Take advantage of precast concrete columns paired with architectural panels to create large and open spaces.

- Multi-Unit Housing: Precast concrete products have superior fire resistance — known to reduce fire insurance rates — and also act as a sound barrier. These characteristics make it a perfect choice for hotels, dormitories, apartment buildings and complexes, senior living communities and similar structures. Shannondell Senior Living is an example of a precast concrete senior living community.

- Hospitals and Medical Centers: For many of the same reasons precast concrete is preferred for multi-unit housing, it also provides a strong foundation for hospitals and medical centers. Hershey Medical Center is an example of a precast concrete hospital building.

- Schools: Precast concrete makes school construction a breeze. With faster turnaround times from start to finish, precast concrete will keep your project on target. Whether you're adding on to a university campus or an elementary school, you'll get students moved in quicker without all the headaches of traditional building. The Charter Arts School is an example of a precast concrete school building.

- Retail Shopping Centers: Retail shopping centers vary — in rural areas, they may be built on a large plot of land, while urban areas tend to have smaller construction sites. They may or may not incorporate parking and can come in single stories, or a few stories high. Regardless of the application, precast concrete has the versatility to match, and it's often used in constructing retail shopping centers. The Target Retail Center is an example of a precast retail shopping building.

Benefits of Precast Concrete

For projects residential and commercial, precast concrete means engineers enjoy greater latitude in planning and design. Precast concrete products arrive on site completely customized and ready for fast installation. When you choose precast concrete products, you can accelerate your project's schedule and enjoy the cost savings that emerge from using concrete products that are precast offsite.

However, the benefits extend beyond convenience and workflow to include versatility, control, efficiency and sustainability, all of which come together to create a superior precast concrete product. Here is some additional information on how precast concrete makes an impact in each of these areas.

1. Versatility

We briefly covered the versatility of precast concrete products already but highlight it again, as it is one of the main benefits of this type of concrete construction. The ability to shape this concrete to include the colors, texture and size you want is a big part of the reason it has implications in such a wide range of industries and uses. While the perception of some is that precast concrete has a lack of versatility, the opposite is true.

2. Controlled Environment

Precast concrete is created in a fully controlled environment, which eliminates any chance of outside variables, like the weather, interfering with the quality or timeline of production. You have complete control over the climate to ensure that all precast concrete products are cured consistently in ideal conditions. Since you have access to these ideal conditions, you can also be confident that production of these precast concrete materials will be on time, as weather delays for pouring on site become a thing of the past.

3. Efficiency

Efficiency comes as a result of the controlled manufacturing environment. When you're able to produce precast concrete all year long in a plant setting, it speeds up the overall construction process. Suddenly you don't need to worry about scheduling pouring concrete for a small window of opportunity on-site, during which weather could postpone your entire project.

Rather, the process is now so efficient that you could even have the precast materials made in advance and store them until you need to get them in place on site. It also saves you time and money because of the assembly line techniques, which require less labor and reduce the stress of coordinating on-site skilled labor and logistics.

4. Sustainability

Precast concrete manufacturing is a sustainable process. In fact, many sustainable building developers use precast concrete for LEED certification. Here's why:

- Concrete is made up of natural aggregates — gravel, sand, rock and water.

- Water used in the process of making concrete is recycled.

- Precast concrete's thermal mass absorbs and releases heat slowly which translates to long-term energy savings.

- Factory environments greatly reduce waste from bracing and formwork, excessive concrete, packaging and debris that build up on-site when you cast in-place.

- Precast structures use less material than products cast on-site. Less raw material is harvested from the environment, and less needs to be disposed of at the end of a building's lifecycle.

- Factory environments are healthier for employees than construction sites because safety hazards, noise and air quality can be controlled.

- Many precast concrete ingredients are produced locally, with aggregates mined a short distance from production, which cuts down on hauling trips.

Roman Concrete

Roman concrete was used from the late Roman Republic until the end of the Roman Empire. It was used to build monuments, large buildings and infrastructure such as roads and bridges. The quality of the concrete was excellent and the buildings and monuments still standing today are a testament to the strength of their construction.

Concrete was usually covered as concrete walls were considered unaesthetic. Roman builders covered building walls with stones or small square tuff blocks that would often form beautiful patterns noting that brick faced concrete buildings were common in Rome especially after the great fire of 64 AD.

Roman Concrete Formula

Roman concrete or opus caementicium was invented in the late 3rd century BC when builders added a volcanic dust called pozzolana to mortar made of a mixture of lime or gypsum, brick or rock pieces and water. Concrete was made by mixing with water:

1) An aggregate which included pieces or rock, ceramic tile, pieces of brick from previously demolished constructions,

2) Volcanic dust (called pozzolana), and

3) Gypsum or lime. Usually the mix was a ratio of 1 part of lime for 3 parts of volcanic ash.

Pozzolana contained both silica and alumina and created a chemical reaction which strengthened the cohesiveness of the mortar.

There were many variations of concrete and Rome even saw the Concrete Revolution which represented advances in the composition of concrete and allowed for the construction of impressive monuments such as the Pantheon. For example, Roman builders discovered that adding crushed terracotta to the mortar created a waterproof material which could be then be used with cisterns and other constructions exposed to rain or water.

Romans mastered underwater concrete by the middle of the 1st century AD. The city of Caesarea gives us an impressive example of Roman construction. The production technique was quite incredible: the mix was one-part lime for two-parts volcanic ash, and it was placed in volcanic tuff or small wooden cases. The seawater would then hydrate the lime and trigger a hot chemical reaction which hardened the concrete.

Caesarea harbor before and today.

Roman Concrete vs Modern Concrete

Actually it has been argued that the concrete used by the Romans was of better quality than the concrete in use today. Recent research from US and Italian scientists has shown that the concrete used to make Roman harbors in the Mediterranean was more resistant than modern concrete (known as Portland cement).

The production process was dramatically different. Portland cement is made by heating clays and limestone at high temperatures (various additives are also added) while the Romans used volcanic ash and a much smaller amount of lime heated at lower temperatures than modern methods.

For example, Roman harbors remain intact today after 2,000 years of waves breaking on the harbors' breakwaters whereas Portland concrete begins to erode in less than 50 years of sea battering. The concrete from ancient Rome also had bending properties that Portland concrete does not have due to its lime and volcanic ash, which explains why it does not crack after a few decades.

Fiber-reinforced Concrete

Fiber Reinforced Concrete can be defined as a composite material consisting of mixtures of cement, mortar or concrete and discontinuous, discrete, uniformly dispersed suitable fibers. Fiber reinforced concrete are of different types and properties with many advantages. Continuous meshes, woven fabrics and long wires or rods are not considered to be discrete fibers.

Fiber is a small piece of reinforcing material possessing certain characteristics properties. They can be circular or flat. The fiber is often described by a convenient parameter called "aspect ratio". The aspect ratio of the fiber is the ratio of its length to its diameter. Typical aspect ratio ranges from 30 to 150.

Fiber reinforced concrete (FRC) is concrete containing fibrous material which increases its structural integrity. It contains short discrete fibers that are uniformly distributed and randomly oriented. Fibers include steel fibers, glass fibers, synthetic fibers and natural fibers. Within these different fibers that character of fiber reinforced concrete changes with varying concretes, fiber materials, geometries, distribution, orientation and densities.

Fibre-reinforcement is mainly used in shotcrete, but can also be used in normal concrete. Fibre-reinforced normal concrete are mostly used for on-ground floors and pavements, but can be considered for a wide range of construction parts (beams, pliers, foundations etc) either alone or with hand-tied rebars.

Concrete reinforced with fibres (which are usually steel, glass or "plastic" fibres) is less expensive than hand-tied rebar, while still increasing the tensile strength many times. Shape, dimension and length of fibre is important. A thin and short fibre, for example short hair-shaped glass fibre, will only be effective the first hours after pouring the concrete (reduces cracking while the concrete is stiffening) but will not increase the concrete tensile strength

Effect of Fibers in Concrete

Fibres are usually used in concrete to control plastic shrinkage cracking and drying shrinkage cracking. They also lower the permeability of concrete and thus reduce bleeding of water. Some types of fibres produce greater impact, abrasion and shatter resistance in concrete. Generally fibres do not increase the flexural strength of concrete, so it can not replace moment resisting or structural steel reinforcement. Some fibres reduce the strength of concrete.

The amount of fibres added to a concrete mix is measured as a percentage of the total volume of the composite (concrete and fibres) termed volume fraction (V_f). V_f typically ranges from 0.1 to 3%. Aspect ratio (l/d) is calculated by dividing fibre length (l) by its diameter. Fibres with a non-circular cross section use an equivalent diameter for the calculation of aspect ratio.

If the modulus of elasticity of the fibre is higher than the matrix (concrete or mortar binder), they help to carry the load by increasing the tensile strength of the material. Increase in the aspect ratio of the fibre usually segments the flexural strength and toughness of the matrix. However, fibres which are too long tend to "ball" in the mix and create workability problems.

Some recent research indicated that using fibres in concrete has limited effect on the impact resistance of concrete materials. This finding is very important since traditionally people think the ductility increases when concrete reinforced with fibres. The results also pointed out that the micro fibres is better in impact resistance compared with the longer fibres.

Necessity of Fiber Reinforced Concrete

1. It increases the tensile strength of the concrete.

2. It reduces the air voids and water voids the inherent porosity of gel.

3. It increases the durability of the concrete.

4. Fibres such as graphite and glass have excellent resistance to creep, while the same is not true for most resins. Therefore, the orientation and volume of fibres have a significant influence on the creep performance of rebars/tendons.

5. Reinforced concrete itself is a composite material, where the reinforcement acts as the strengthening fibre and the concrete as the matrix. It is therefore imperative that the behavior under thermal stresses for the two materials be similar so that the differential deformations of concrete and the reinforcement are minimized.

6. It has been recognized that the addition of small, closely spaced and uniformly dispersed fibers to concrete would act as crack arrester and would substantially improve its static and dynamic properties.

Factors Affecting Properties of Fiber Reinforced Concrete

Fiber reinforced concrete is the composite material containing fibers in the cement matrix in an orderly manner or randomly distributed manner. Its properties would obviously, depends upon the efficient transfer of stress between matrix and the fibers. The factors are briefly discussed below:

1. Relative Fiber Matrix Stiffness

The modulus of elasticity of matrix must be much lower than that of fiber for efficient stress transfer. Low modulus of fiber such as nylons and polypropylene are, therefore, unlikely to give strength improvement, but the help in the absorption of large energy and therefore, impart greater degree of toughness and resistance to impart. High modulus fibers such as steel, glass and carbon impart strength and stiffness to the composite.

Interfacial bond between the matrix and the fiber also determine the effectiveness of stress transfer, from the matrix to the fiber. A good bond is essential for improving tensile strength of the composite.

2. Volume of Fibers

The strength of the composite largely depends on the quantity of fibers used in it. Figure shows the effect of volume on the toughness and strength. It can be seen from Figure that the increase in the volume of fibers, increase approximately linearly, the tensile strength and toughness of the composite. Use of higher percentage of fiber is likely to cause segregation and harshness of concrete and mortar.

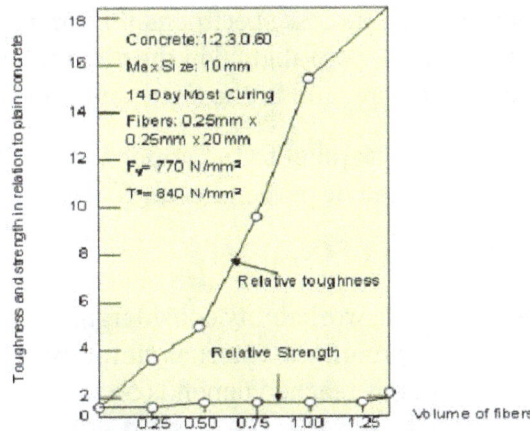

Figure: Effect of volume of fibers in flexure.

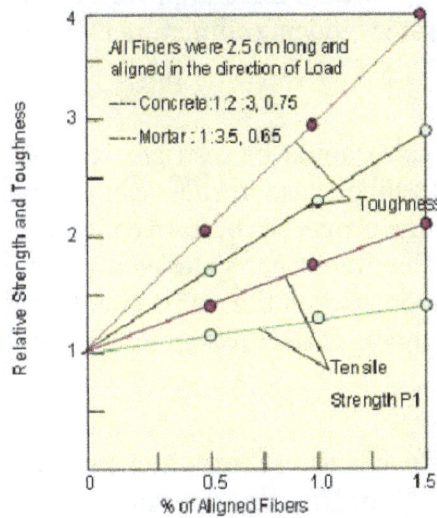

Figure: Effect of volume of fibers in tension.

3. Aspect Ratio of the Fiber

Another important factor, which influences the properties and behavior of the composite, is the aspect ratio of the fiber. It has been reported that up to aspect ratio of 75, increase on the aspect ratio increases the ultimate concrete linearly. Beyond 75, relative strength and toughness is reduced. Table below shows the effect of aspect ratio on strength and toughness.

Type of concrete	Aspect ratio	Relative strength	Relative toughness
Plain concrete	0	1	1
With	25	1.5	2.0
Randomly	50	1.6	8.0
Dispersed fibers	75	1.7	10.5
	100	1.5	8.5
Table: Aspect Ratio of the Fiber			

4. Orientation of Fibers

One of the differences between conventional reinforcement and fiber reinforcement is that in conventional reinforcement, bars are oriented in the direction desired while fibers are randomly oriented. To see the effect of randomness, mortar specimens reinforced with 0.5% volume of fibers were tested. In one set specimens, fibers were aligned in the direction of the load, in another in the direction perpendicular to that of the load, and in the third randomly distributed.

It was observed that the fibers aligned parallel to the applied load offered more tensile strength and toughness than randomly distributed or perpendicular fibers.

5. Workability and Compaction of Concrete

Incorporation of steel fiber decreases the workability considerably. This situation adversely affects the consolidation of fresh mix. Even prolonged external vibration fails to compact the concrete. The fiber volume at which this situation is reached depends on the length and diameter of the fiber.

Another consequence of poor workability is non-uniform distribution of the fibers. Generally, the workability and compaction standard of the mix is improved through increased water/ cement ratio or by the use of some kind of water reducing admixtures.

6. Size of Coarse Aggregate

Maximum size of the coarse aggregate should be restricted to 10mm, to avoid appreciable reduction in strength of the composite. Fibers also in effect, act as aggregate. Although they have a simple geometry, their influence on the properties of fresh concrete is complex. The inter-particle friction between fibers and between fibers and aggregates controls the orientation and distribution of the fibers and consequently the properties of the composite. Friction reducing admixtures and admixtures that improve the cohesiveness of the mix can significantly improve the mix.

7. Mixing

Mixing of fiber reinforced concrete needs careful conditions to avoid balling of fibers, segregation and in general the difficulty of mixing the materials uniformly. Increase in the aspect ratio, volume percentage and size and quantity of coarse aggregate intensify the difficulties and balling tendency. Steel fiber content in excess of 2% by volume and aspect ratio of more than 100 are difficult to mix.

It is important that the fibers are dispersed uniformly throughout the mix; this can be done by the addition of the fibers before the water is added. When mixing in a laboratory mixer, introducing the fibers through a wire mesh basket will help even distribution of fibers. For field use, other suitable methods must be adopted.

Different Types of Fiber Reinforced Concrete

Following are the different type of fibers generally used in the construction industries:

1. Steel Fiber Reinforced Concrete

A no of steel fiber types are available as reinforcement. Round steel fiber the commonly used type are produced by cutting round wire in to short length. The typical diameter lies in the range of 0.25 to 0.75 mm. Steel fibers having a rectangular c/s are produced by silting the sheets about 0.25 mm thick.

Fiber made from mild steel drawn wire. Conforming to IS:280-1976 with the diameter of wire varying from 0.3 to 0.5 mm have been practically used in India.

Round steel fibers are produced by cutting or chopping the wire, flat sheet fibers having a typical c/s ranging from 0.15 to 0.41 mm in thickness and 0.25 to 0.90 mm in width are produced by silting flat sheets.

Deformed fiber, which are loosely bounded with water-soluble glue in the form of a bundle are also available. Since individual fibers tend to cluster together, their uniform distribution in the matrix is often difficult. This may be avoided by adding fibers bundles, which separate during the mixing process.

2. Polypropylene Fiber Reinforced (PFR) cement mortar and concrete

Polypropylene is one of the cheapest & abundantly available polymers polypropylene fibers are resistant to most chemical & it would be cementitious matrix which would deteriorate first under aggressive chemical attack. Its melting point is high (about 165 degrees centigrade). So that a working temp. As (100 degree centigrade) may be sustained for short periods without detriment to fiber properties.

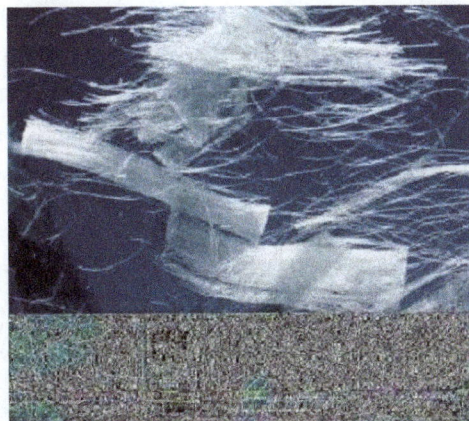

Figure: Polypropylene fiber reinforced cement-mortar and concrete.

Polypropylene fibers being hydrophobic can be easily mixed as they do not need lengthy contact during mixing and only need to be evenly distressed in the mix.

Polypropylene short fibers in small volume fractions between 0.5 to 15 commercially used in concrete.

3. GFRC – Glass Fiber Reinforced Concrete

Glass fiber is made up from 200-400 individual filaments which are lightly bonded to make up a stand. These stands can be chopped into various lengths, or combined to make cloth mat or tape. Using the conventional mixing techniques for normal concrete it is not possible to mix more than about 2% (by volume) of fibers of a length of 25 mm.

The major appliance of glass fiber has been in reinforcing the cement or mortar matrices used in the production of thin-sheet products. The commonly used verities of glass fibers are e-glass used. In the reinforced of plastics & AR glass E-glass has inadequate resistance to alkalis present in Portland cement where AR-glass has improved alkali resistant characteristics. Sometimes polymers are also added in the mixes to improve some physical properties such as moisture movement.

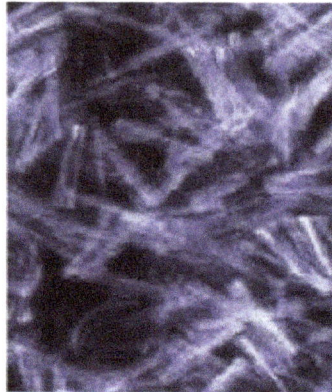
Figure: Glass-fiber reinforced concrete.

4. Asbestos Fibers

The naturally available inexpensive mineral fiber, asbestos, has been successfully combined with Portland cement paste to form a widely used product called asbestos cement. Asbestos fibers here thermal mechanical & chemical resistance making them suitable for sheet product pipes, tiles and corrugated roofing elements. Asbestos cement board is approximately two or four times that of unreinforced matrix. However, due to relatively short length (10 mm) the fiber have low impact strength.

Figure: Asbestos fiber.

5. Carbon Fibers

Carbon fibers from the most recent & probability the most spectacular addition to the range of fiber available for commercial use. Carbon fiber comes under the very high modulus of elasticity and flexural strength. These are expansive. Their strength & stiffness characteristics have been found to be superior even to those of steel. But they are more vulnerable to damage than even glass fiber, and hence are generally treated with resign coating.

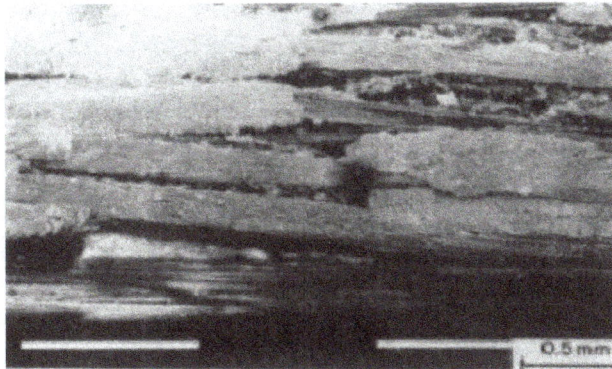
Figure: Carbon fibers.

6. Organic Fibers

Organic fiber such as polypropylene or natural fiber may be chemically more inert than either steel or glass fibers. They are also cheaper, especially if natural. A large volume of vegetable fiber may be used to obtain a multiple cracking composite. The problem of mixing and uniform dispersion may be solved by adding a superplasticizer.

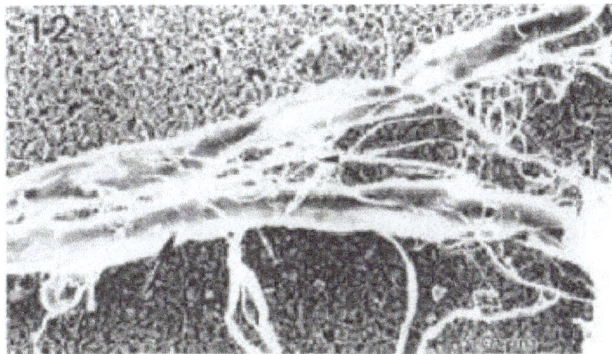
Figure: Organic fibe.

Autoclaved Aerated Concrete

Autoclaved aerated concrete is a versatile lightweight construction material and usually used as blocks. Compared with normal (ie: "dense" concrete) aircrete has a low density and excellent insulation properties.

The low density is achieved by the formation of air voids to produce a cellular structure. These voids are typically 1mm-5mm across and give the material its characteristic appearance. Blocks typically have strengths ranging from 3-9 Nmm^{-2} (when tested in accordance with BS EN 771-1:2000). Densities range from about 460 to 750 kg m^{-3}; for comparison, medium density concrete blocks have a typical density range of 1350-1500 kg m^{-3} and dense concrete blocks a range of 2300-2500 kg m^{-3}.

Figure: Autoclaved aerated concrete block with
a sawn surface to show the cellular pore structure.

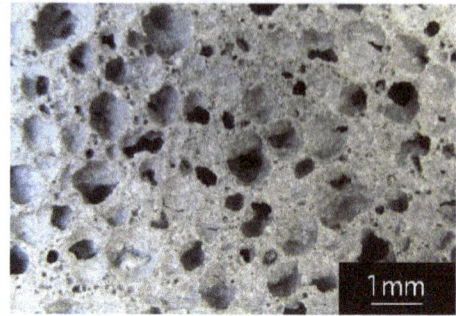

Figure: Detailed view of cellular pore structure
in an aircrete block.

Autoclaved aerated concrete blocks are excellent thermal insulators and are typically used to form the inner leaf of a cavity wall. They are also used in the outer leaf, when they are usually rendered, and in foundations. It is possible to construct virtually an entire house from autoclaved aerated concrete, including walls, floors - using reinforced aircrete beams, ceilings and the roof. Autoclaved aerated concrete is easily cut to any required shape.

Aircrete also has good acoustic properties and it is durable, with good resistance to sulfate attack and to damage by fire and frost.

Production

Autoclaved aerated concrete is cured in an autoclave - a large pressure vessel. In aircrete production the autoclave is normally a steel tube some 3 metres in diameter and 45 metres long. Steam is fed into the autoclave at high pressure, typically reaching a pressure of 800 kPa and a temperature of 180° C.

Autoclaved aerated concrete can be produced using a wide range of cementitous materials, commonly:

- Portland cement, lime and pulverised fuel ash (PFA, fly ash)

or

- Portland cement, lime and fine silica sand. The sand is usually milled to achieve adequate fineness.

A small amount of anhydrite or gypsum is also often added.

Autoclaved aerated concrete is quite different from dense concrete (i.e. "normal concrete") in both the way it is produced and in the composition of the final product.

Dense concrete is typically a mixture of cement and water, often with slag or PFA, and fine and coarse aggregate. It gains strength as the cement hydrates, reaching 50% of its final strength after perhaps about 2 days and most of its final strength after a month.

In contrast, autoclaved aerated concrete is of much lower density than dense concrete. The chemical reactions forming the hydration products go virtually to completion during autoclaving and so when removed from the autoclave and cooled, the blocks are ready for use.

Autoclaved aerated concrete does not contain any aggregate; all the main mix components are reactive, even milled sand where it is used. The sand, inert when used in dense concrete, behaves as a pozzolan in the autoclave due to the high temperature and pressure.

The autoclaved aerated concrete production process differs slightly between individual production plants but the principles are similar. We will assume a mix that contains cement, lime and sand; these are mixed to form slurry. Also present in the slurry is fine aluminum powder - this is added to produce the cellular structure. The density of the final block can be varied by changing the amount of aluminum powder in the mix.

The slurry is poured into molds that resemble small railway wagons with drop-down sides. Over a period of several hours, two processes occur simultaneously:

- The cement hydrates normally to produce ettringite and calcium silicate hydrates and the mix gradually stiffens to form what is termed a "green cake".

- The green cake rises in the mold due to the evolution of hydrogen gas formed from the reaction between the fine aluminum particles and the alkaline liquid. These gas bubbles give the material its cellular structure.

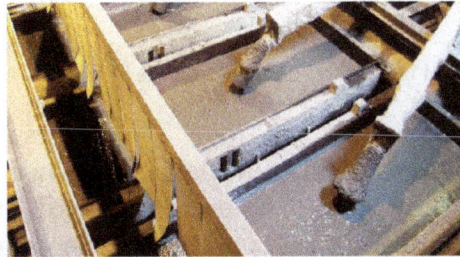

Figure: Slurry being poured into molds.

At the risk of incurring the ire of aircrete manufacturers, it could be said that there are parallels between autoclaved aerated concrete production and bread-making. In bread, the dough contains yeast and is mixed, then left to rise as the yeast converts sugars to carbon dioxide.

The dough must have the right consistency; too hard and the bubbles of carbon dioxide cannot 'stretch' the dough to make it rise, but if the dough is too sloppy, the carbon dioxide bubbles rise to the surface and are lost and the dough collapses. With the right consistency, the dough is sufficiently elastic to stretch and expand, but strong enough to retain the gas so that the dough does not collapse. When risen, the dough is placed in the oven.

Figure: Green cake rising in mold.

Although a much more complex process, Aircrete production conditions are precisely-controlled for, in part, somewhat similar reasons. The mix proportions and the initial mix temperature must be correct and the aluminum powder must be present in the required amount and with the appropriate reactivity an an alkaline environment. All of the materials be be of suitable fineness. A complicating factor is that the temperature of the green cake increases due to the exothermic reactions as the lime and the cement hydrate, so the reactions proceed faster.

When the cake has risen to the required height, the mold moves along a track to where the cake is cut to the required block size. Depending on the actual production process, the cake may be demolded entirely onto a trolley before cutting, or it may be cut in the mold after the sides are removed.

The cake is cut by passing through a series of cutting wires.

Figure: Green cake being cut by wires.

At the cutting stage, the blocks are still green - only a few hours have passed since the mix was poured into the mold and they are soft and easily damaged. However, if they are too soft, the cut blocks may either fall apart or stick together; if they are too hard, the wires will not cut them here too, the process has to be carefully controlled to achieve the necessary consistency.

The cut blocks are then loaded into the autoclave. It takes a couple of hours for the autoclave to reach maximum temperature and pressure, which is held for perhaps 8-10 hours, or longer for high density/high strength aircrete.

Figure: "Green" blocks being loaded into an autoclave.

When removed from the autoclave and cooled, the blocks have achieved their full strength and are packed ready for transport.

AAC Composition

The essence of aircrete production is that lime from the cement and lime in the mix reacts with silica to form 1.1 nm tobermorite.

cement

Lime + Silica ⟹ **1.1nm tobermorite**

↑   ↑

lime   *sand*

Figure: Components and products in aircrete production.

During the green stage, the cement is hydrating at normal temperatures and the hydration products are initially similar to those in dense concrete - C-S-H, CH and ettringite and/or monosulfate. After autoclaving, tobermorite is normally the principal final reaction product due to the high temperature and pressure.

Small amounts of other hydrated phases will also be present in the final product. Additionally, hydrated phases form in the autoclave as intermediate products, principally C-S-H(I). This is a more crystalline form of calcium silicate hydrate than occurs in dense concrete; it can have a ratio of calcium to silicon of ($0.8 < Ca/Si < 1.5$) but 0.8 to 1.0 is desirable as this ratio facilitates the formation of 1.1 nm tobermorite.

The compositions of the hydration products in aircrete are therefore quite different from those in dense concrete cured at normal temperatures (ie: calcium silicate hydrate (C-S-H), calcium hydroxide (CH), ettringite and monosulfate.

Looking at this in a little more detail from when the green blocks enter the autoclave, the main reactions that occur are broadly as follows:

- Over 2 hours or so, as the pressure and temperature increase, the normal cement hydration products that formed in the green state progressively disappear and the sand becomes reactive.

- C-S-H(I) forms, partly from silica derived from the sand.

- As more sand reacts, calcium hydroxide from the lime and from cement hydration is gradually used up by continued formation of C-S-H(I).

- With continued autoclaving, 1.1 nm tobermorite starts to crystallize from the C-S-H(I); the total proportion of C-S-H(I) declines and that of 1.1 nm tobermorite gradually increases. C-S-H(I) is therefore mainly an intermediate compound.

The final hydration products are then principally:

- 1.1 nm tobermorite

- Possibly some residual C-S-H(I)

- Hydrogarnet

Unreacted sand is likely to remain in the final product. There may also be some residual calcium hydroxide if insufficient silica has reacted and some residual anhydrite and/or hydroxyl-ellestadite if anhydrite was present in the mix.

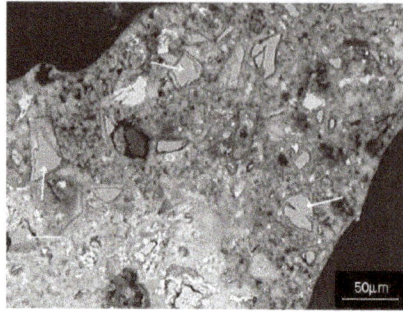

Figure: SEM image of polished section showing a detail -
a cell wall - of a block made with cement, lime and sand mix.

In Figure, some residual unreacted sand particles remain, often with rims of hydration product showing the size of the original particle. Most of the matrix is composed of tobermorite. Black areas at top left and bottom right are epoxy resin used in preparing the polished section filling air voids (air cells).

The objective is to react sufficient silica from the sand to form tobermorite from the available lime supplied by the lime and cement. This will depend on a range of factors, including the inherent reactivities of the materials, their fineness (especially the sand), and the temperature and pressure. If the autoclaving time is too short, the tobermorite content will not be maximised and some un-reacted calcium hydroxide will remain and block strengths will be then less than optimum. If the autoclaving time is too long, other hydration products may form which may also be detrimental to strength and an unnecessary energy cost will be incurred.

There are different forms of tobermorite: 1.1 nm tobermorite and 1.4 nm tobermorite. Also, there are different types of 1.1 nm tobermorite and these behave differently when heated. Their crystal structure is that of layered sheets, with water molecules between the layers - on heating, the inter-layer water is lost; as a result, some 1.1 nm tobermorites shrink (a process known as lattice shrinkage) but some don't.

1.4 nm tobermorite $(C_5S_6H_9)$ - forms at room temperature and is found as a natural mineral. It decomposes at 55° C to 1.1 nm tobermorite and so is not found in AAC.

Calcium Silicate Hydrate Compositions in AAC

- 1.1nm tobermorite $(C_5S_6H_5)$ is usually the main hydration product in AAC where cement, lime and sand are used

- C-S-H(I) - more crystalline than C-S-H in dense concrete, typically $0.8 < Ca/Si < 1.0$

- Xonotlite (C_6S_6H) - forms with longer autoclaving times, or higher temperatures

'Normal' tobermorite shows lattice shrinkage, while non-shrinking tobermorite is called 'anomalous' tobermorite. Tobermorite in AAC made with cement, lime and sand is usually normal tobermorite. Tobermorite in autoclaved aerated concrete made with cement, lime and PFA is usually anomalous tobermorite. Aluminium and alkali together in solution (such as will be present in mixes of cement, lime and PFA) tend to produce anomalous tobermorite, with some aluminium and alkali taken up into the tobermorite crystal structure. The differences between the different forms of autoclaved calcium silicate hydrates are not well-defined; in an AAC block, intimately-mixed hydrates of different compositions and crystallinity are likely to occur.

Other Hydrothermally-formed Minerals:

- Gyrolite ($C_2S_3H_2$) - not normally found in AAC

- Jennite ($C_9S_6H_{11}$) occurs as a natural mineral; not found in AAC

- C-S-H(II) - Ca/Si $\approx$ 2.0. Does not occur in AAC

- C_2SH (α-C_2S hydrate) can occur in autoclaved products but is undesirable

- Hydroxyl-ellestadite ($C_{10}S_3.3SO_3.H_2O$) - may be found in AAC; also occurs at the cooler end of cement kilns

Performance

Appearance

Autoclaved aerated concrete is light coloured. It contains many small voids (similar to those in aerated chocolate bars) that can be clearly seen when looked at closely. The gas used to 'foam' the concrete during manufacture is hydrogen formed from the reaction of aluminium paste with alkaline elements in the cement. These air pockets contribute to the material's insulating properties. Unlike masonry, there is no direct path for water to pass through the material; however, it can wick up moisture and an appropriate coating is required to prevent water penetration.

AAC used in veneer construction.

Structural Capability

The compressive strength of AAC is very good. Although it is one-fifth the density of normal concrete it still has half the bearing strength, and loadbearing structures up to three storeys high can be safely erected with AAC blockwork. Increasingly, AAC is being used in Australia in its panel form as a cladding system rather than as a loadbearing wall. Entire building structures can be made in AAC from walls to floors and roofing with reinforced lintels, blocks and floor, wall and roofing panels available from the manufacturer.

The Australian Standard AS 3700-2011, Masonry structures, includes provisions for AAC block design. External AAC wall panels — which are not blockwork but are precast units — can provide loadbearing support in houses up to two storeys high. AAC panels and lintels contain integral steel reinforcement to ensure structural adequacy during installation and design life.

AAC floor panels can be used to make non-loadbearing concrete floors that can be installed by carpenters.

Block construction in a two storey house.

Thermal Mass

The thermal mass performance of AAC is dependent on the climate in which it is used. With its mixture of concrete and air pockets, AAC has a moderate overall level of thermal mass performance. Its use for internal walls and flooring can provide significant thermal mass. The temperature moderating thermal mass is most useful in climates with high cooling needs.

Insulation

AAC has very good thermal insulation qualities relative to other masonry but generally needs additional insulation.

A 200 mm thick AAC wall gives an R-value rating of 1.43 with 5% moisture content by weight. With a 2–3 mm texture coating and 10mm plasterboard internal lining it achieves an R rating of 1.75 (a cavity brick wall achieves 0.82). The BCA requires that external walls in most climate zones must achieve a minimum total R-value of 2.8.

To comply with building code provisions for thermal performance, a 200 mm AAC blockwork wall requires additional insulation.

AAC panels on a lightweight timber framed house.

A texture-coated 100mm AAC veneer on a lightweight 70 mm or 90 mm frame filled with bulk insulation achieves a higher R rating than an otherwise equivalent brick veneer wall.

Relative to their thickness, AAC panels provide less insulation than AAC blockwork, e.g. a 100 mm blockwork AAC wall has a dry state R-value of 0.86 and a 100mm AAC wall panel has a dry state R-value of 0.68.

Loadbearing, insulating and capable of being sculpted, AAC has enormous potential as an environmentally responsible choice of building material.

Sound Insulation

With its closed air pockets, AAC can provide very good sound insulation. As with all masonry construction, care must be taken to avoid gaps and unfilled joints that can allow unwanted sound transmission. Combining the AAC wall with an insulated asymmetric cavity system gives wall excellent sound insulation properties.

Fire and Vermin Resistance

AAC is inorganic, incombustible and does not explode; it is thus well suited for fire-rated applications. Depending on the application and the thickness of the blocks or panels, fire ratings up to four hours can be achieved. AAC does not harbour or encourage vermin.

Durability and Moisture Resistance

The purposely lightweight nature of AAC makes it prone to impact damage. With the surface protected to resist moisture penetration it is not affected by harsh climatic conditions and does not degrade under normal atmospheric conditions. The level of maintenance required by the material varies with the type of finish applied.

The porous nature of AAC can allow moisture to penetrate to a depth but appropriate design (damp proof course layers and appropriate coating systems) prevents this happening. AAC does not easily degrade structurally when exposed to moisture, but its thermal performance may suffer.

A number of proprietary finishes (including acrylic polymer based texture coatings) give durable and water resistant coatings to AAC blockwork and panels. They need to be treated in a similar fashion with acrylic polymer based coatings before tiling in wet areas such as showers. The manufacturer can advise on the appropriate coating system, surface preparation and installation instructions to give good water repellent properties.

Plasticised, thin coat finishes are common, but here a non-plasticised thick coat render (10 mm approximately) was used. Some variation in the amount of show-through of the blockwork pattern can be seen in this example, which also illustrates the use of glass blocks as well as more conventional windows.

Toxicity and Breathability

The aerated nature of AAC facilitates breathability. There are no toxic substances and no odour in the final product. However, AAC is a concrete product and calls for precautions similar to those for handling and cutting concrete products. It is advisable to wear personal protective equipment such as gloves, eye wear and respiratory masks during cutting, due to the fine dust produced by concrete products. If low-toxic, vapour permeable coatings are used on the walls and care is taken not to trap moisture where it can condense, AAC may be an ideal material for homes for the chemically sensitive.

Autoclaved aerated concrete is about one-fifth the density of normal concrete blocks.

Environmental Impacts

Weight for weight, AAC has manufacturing, embodied energy and greenhouse gas emission impacts similar to those of concrete, but can be up to one-quarter to one-fifth that of concrete based on volume. AAC products or building solutions may have lower embodied energy per square metre than a concrete alternative. In addition, AAC's much higher insulation value reduces heating and cooling energy consumption. AAC has some significant environmental advantages over conventional construction materials, addressing longevity, insulation and structural demands in one material. As an energy and material investment it can often be justified for buildings intended to have a long life.

Offcuts from construction can be returned to the manufacturer for recycling, or be sent out as concrete waste for reuse in aggregates; alternatively, the odd pieces can be used directly for making, for example, garden walls or landscape features.

The clear difference between the lower and higher course of blockwork in the loadbearing AAC walls of an apartment building under construction shows the difference in quality that can be achieved with the same material by differently skilled tradespeople.

Buildability, Availability and Cost

Although AAC is relatively easy to work, is one-fifth the weight of concrete, comes in a variety of sizes and is easily carved, cut and sculpted, it nevertheless requires careful and accurate placement: skilled trades and good supervision are essential. Competent bricklayers or carpenters can work successfully with AAC but dimensional tolerances are very small when blockwork is laid with thin-bed mortar. Thick-bed mortar is more forgiving but is uncommon and not the industry preferred option. Very large block sizes may require two-handed lifting and be awkward to handle but can result in fewer joints and more rapid construction.

The construction process with AAC produces little waste as blockwork offcuts can be reused in wall construction. Good design that responds to the regime of standardised panel sizes encourages low-waste, resource-efficient AAC panel construction.

The cost of AAC is moderate to high. In Australia, AAC is competitive with other masonry construction but more expensive than timber frame. Lack of competition in the marketplace makes consumers highly dependent on one manufacturer.

This isometric concept shows the versatility of AAC products in dwelling construction.

Typical Domestic Construction

Construction Process

All structural design should be prepared by a competent person, and may require preparation and approval by a qualified engineer. Qualified professionals, architects and designers bring years of experience and access to intellectual property that has the potential to save house builders time and money, and help achieve environmental performance objectives. All masonry construction has to comply with the BCA and relevant Australian Standards, e.g. all masonry walls are required to have movement or expansion joints at specified intervals.

The standard block size is 200 mm high by 600 mm long. Block thickness can range from 50 mm to 300 mm but for residential construction the most common block widths used are 100 mm, 150 mm and 200 mm. AAC blocks can be used in a similar manner to traditional masonry units such as bricks: they can be applied as a veneer in timber frame or serve as one or both skins in cavity wall construction.

The standard panel size is 600 mm wide by 75 mm thick with lengths ranging from 1200 mm to 3000 mm. AAC panels can be used as a veneer cladding over timber or steel-framed construction.

The AAC manufacturer provides a wealth of detailed technical advice that, if followed, should help to ensure successful use of the product.

Movement Joints

Movement joints must be provided at 6 m horizontal centres maximum (measured continuously around rigid corners).

Footings

AAC block construction requires level footings designed for full or articulated masonry in accordance with AS 2870-2011, Residential slabs and footings. Stiff footings are preferred because the wall structure of thin-bed mortar AAC acts as if it were a continuous material and cracking tends not to follow the mortar beds and joints the way it does in traditional masonry walling. Thick-bed mortar AAC walls act more like traditional masonry but are not the preferred method for AAC.

Frames

Frames may be required for various structural reasons. Earthquake provisions tend to require multi-storey AAC structures to have a frame of steel or reinforcement to withstand potential earthquake loads that may induce strong, sharp horizontal forces. It is a relatively simple matter to build AAC blockwork around steel frames but embedding reinforcement rods can be costly and difficult.

AAC panels on lightweight steel framed houses.

Joints and Connections

The AAC manufacturer provides proprietary mortar mixes. Although more conventional thick-bed mortar (approximately 10 mm) can be used with AAC, the manufacturer's approved option is a proprietary thin-bed mortar. With this method, the procedure of laying the blocks is more like gluing than conventional brickwork construction. This is why many traditionally trained bricklayers may need some time to adjust to this different method of working. In addition, brickies are used to lifting bricks with a single hand and AAC blocks often require two-handed manipulation. Although this may appear a slower construction process than laying masonry units, an AAC block is equivalent to five or six standard bricks.

Loadbearing Walls

AAC is available in blocks of various sizes and in larger reinforced panels, sold as part of a complete building system that includes floor and roof panels, and interior and exterior walls.

Fixings

AAC has low compression strength. The use of mechanical fasteners is not recommended, as repeated loading of the fastener can result in local crushing of the AAC and loosening of the fastener. Proprietary fasteners are specifically designed to accommodate the nature of the material by spreading the forces created by any given load, whether it is a beam, shelf or picture hook. A number of proprietary fixings for AAC come with extensive guidance in product literature. If you are not sure, consult the project engineer or fastener manufacturer for guidance.

Openings

AAC is soft enough to be cut with hand tools. Niches can be carved into thicker walls, corners can be chamfered or curved for visual effect and you can easily make channels for pipes and wires with an electric router. Use appropriate dust reduction strategies with all carving and cutting, and wear appropriate personal protection equipment at all times.

Dry-lined interior shows how AAC can be exploited to make niches and unusual openings.

Finishes

AAC blockwork and panels can accept cement render, but the manufacturer recommends using a proprietary render mix compatible with the AAC material substrate. Site-mixed cement renders have to be compatible with the AAC substrate, with the render having a lower strength than conventional renders. All renders should be vapour permeable (but water-resistant) to achieve a healthy breathable construction. All external-coating finishes should provide good UV resistance, be vapour permeable and be proven suitable for AAC.

Polished Concrete

A polished concrete floor has a glossy, mirror-like finish. The design options for polished concrete are vast. We can choose nearly any color, create patterns with saw cuts, or embed aggregates or interesting objects into the concrete prior to polishing. The reflectivity of the floor can also be controlled by using different levels of concrete polishing. Polished concrete is popular in commercial buildings because it is easy to maintain. Maintaining polished floors requires dust mopping and occasional use of a cleaning product.

Simply put, polishing concrete is similar to sanding wood. Heavy-duty polishing machines equipped with progressively finer grits of diamond-impregnated segments or disks (akin to sandpaper) are used to gradually grind down surfaces to the desired degree of shine and smoothness.

HTC Professional Floor Systems in Knoxville, TN

Polishing Process

The process begins with the use of coarse diamond segments bonded in a metallic matrix. These segments are coarse enough to remove minor pits, blemishes, stains, or light coatings from the floor in preparation for final smoothing. Depending on the condition of the concrete, this initial rough grinding is generally a three- to four-step process.

The next steps involve fine grinding of the concrete surface using diamond abrasives embedded in a plastic or resin matrix. Crews use ever-finer grits of polishing disks (a process called lapping) until the floor has the desired sheen. For an extremely high-gloss finish, a final grit of 1500 or finer may be used. Experienced polishing crews know when to switch to the next-finer grit by observing the floor surface and the amount of material being removed.

Decorative Concrete Institute in Temple, GA

During the polishing process an internal impregnating sealer is applied. The sealer sinks into the concrete and is invisible to the naked eye. It not only protects the concrete from the inside out, it also hardens and densifies the concrete. This eliminates the need for a topical coating, which reduces maintenance significantly (versus if we had a coating on it). Some contractors spread a commercial polishing compound onto the surface during the final polishing step, to give the floor a bit more sheen. These compounds also help clean any residue remaining on the surface from the polishing process and leave a dirt-resistant finish.

We can polish concrete using wet or dry methods. Although each has its advantages, dry polishing is the method most commonly used in the industry today because it's faster, more convenient, and environmentally friendly. Wet polishing uses water to cool the diamond abrasives and eliminate grinding dust. Because the water reduces friction and acts as a lubricant, it increases the life of the polishing abrasives. The chief disadvantage of this method is the cleanup. Wet polishing creates a tremendous amount of slurry that crews must collect and dispose of in an environmentally sound manner. With dry polishing, no water is required. Instead, the floor polisher is hooked up to a dust-containment system that vacuum up virtually all of the mess.

Many contractors use a combination of both the wet and dry polishing methods. Typically, dry polishing is used for the initial grinding steps, when more concrete is being removed. As the surface becomes smoother, and crews switch from the metal-bonded to the finer resin-bonded diamond abrasives, they generally change to wet polishing.

Step-by-Step:

- Remove existing coatings (for thick coatings, use a 16- or 20-grit diamond abrasive or more aggressive tool specifically for coating removal, such as a T-Rex).

- Seal cracks and joints with an epoxy or other semi-rigid filler.

- Grind with a 30- or 40-grit metal-bonded diamond.

- Grind with an 80-grit metal-bonded diamond.

- Grind with a 150-grit metal-bonded diamond (or finer, if desired).

- Apply a chemical hardener to densify the concrete.

- Polish with a 100- or 200-grit resin-bond diamond, or a combination of the two.

- Polish with a 400-grit resin-bond diamond.

- Polish with an 800-grit resin-bond diamond.

- Finish with a 1500- or 3000-grit resin-bond diamond (depending on the desired sheen level).

- Optional: Apply a stain guard to help protect the polished surface and make it easier to maintain.

Some of the benefits of polished concrete include:

- Elimination of Dusting from Efflorescence: In ordinary unpolished concrete, tiny particles of dust are pushed to the surface through an upward force called hydrostatic pressure, resulting in efflorescence. Efflorescence leads to dusting, which forces epoxies off of the surface of concrete floors and can make maintenance a costly priority.

- Stain-Resistant: By densifying and sealing the surface, polished concrete transforms a porous concrete floor into a tightened floor that is dense enough to repel water, oil and other contaminants, preventing them from penetrating the surface.

- Improved Reflectivity and Ambient Lighting: The reflective property of a polished concrete floor increases the lighting in facilities. Increased ambient lighting will reduce the energy bill as well as look beautiful.

- Increased Slip Resistance: Polished concrete, though quite shiny, does not create a slippery floor. In fact, the benefits of mechanically grinding and flattening the floor will increase the coefficient of friction when compared to ordinary concrete. Polished concrete often exceeds OSHA standards for floors.

- Less Maintenance: Most floor systems, including tile and linoleum, require aggressive scrubbing to maintain a clean environment and nice appearance. Polished concrete surfaces are tightly compacted, reducing stains and do not require any waxing or stripping to maintain the sheen.

- Cost-Effective: Polished concrete will reduce energy and maintenance costs significantly through reflectivity and ambient lighting, reduction in upkeep (such as waxing) and reduced tire wear.

- LEED Friendly: Polished concrete not only utilizes existing concrete surfaces, eliminating additional materials such as coverings/coatings and moving towards sustainable building, it typically contains no noticeable VOC's, making it friendly for any USBG LEED project.

- Improved condition for old floors: (Mechanical Polish Only) As concrete ages, surface stress, delamination, curled cold joints and other problems can arise. Mechanically grinding the floor will remove the top surface of the old concrete and polishing will then strengthen it, increasing its impact and abrasion resistance.

- Reduced Tire Wear: (Mechanical Polish) The rough, uneven texture of natural concrete causes tires to abrade, adding to their wear. A polished concrete floor system will level the joints and make the entire surface smooth, preventing this abrasion.

- No Production/Plant Shutdowns: (Mechanical Polish) Dry-Mechanically polished concrete can be put into service immediately after the process is complete. Due to the cleanliness of the process and the lack of toxic or hazardous chemicals, floors can often be serviced while the plant is in full production.

Cement

Cement, one of the most important building materials, is a binding agent that sets and hardens to adhere to building units such as stones, bricks, tiles etc. Cement generally refers to a very fine powdery substance chiefly made up of limestone (calcium), sand or clay (silicon), bauxite (aluminum) and iron ore, and may include shells, chalk, marl, shale, clay, blast furnace slag, slate. The raw ingredients are processed in cement manufacturing plants and heated to form a rock-hard substance, which is then ground into a fine powder to be sold. Cement mixed with water causes a chemical reaction and forms a paste that sets and hardens to bind individual structures of building materials.

Cement is an integral part of the urban infrastructure. It is used to make concrete as well as mortar, and to secure the infrastructure by binding the building blocks. Concrete is made of cement, water, sand, and gravel mixed in definite proportions, whereas mortar consists of cement, water, and lime aggregate. These are both used to bind rocks, stones, bricks and other building units, fill or seal any gaps, and to make decorative patterns. Cement mixed with water silicates and aluminates, making a water repellant hardened mass that is used for water-proofing.

Cement, though different from the refined product found nowadays, has been used in many forms since the advent of human civilization. From volcanic ashes, crushed pottery, burnt gypsum and hydrated lime to the first hydraulic cement used by the Romans in the middle ages, the development of cement continued to the 18th century, when James Parker patented Roman cement, which gained popularity but was replaced by Portland cement in the 1850s.

In the 19th century, Frenchman Louis Vicat laid the foundation for the chemical composition of Portland cement and in Russia, Egor Cheliev published the methods of making cement, uses of cement and advantages. Joseph Aspdin brought Portland cement to the market in England and his son, William Aspdin, developed the "modern" Portland cement, which was soon in quite high demand. But the real father of Portland cement is considered to be Isaac Charles Johnson, who contributed immensely by publishing the process of developing meso-Portland cement in the kiln.

In the 19th century, Rosendale cement was discovered in New York. Though its rigidity made it quite popular at first, the market demand soon declined because of its long curing time and Portland cement was again the favorite. However, a new blend of Rosendale-Portland cement, which is both highly durable and needs less curing time, was synthesized by Catskill Aqueduct and is now often used for highway or bridge construction.

The cement used today has undergone experimentation, testing and significant improvements to meet the needs of the present world, such as developing strong concretes for roads and highways, hydraulic mortars that endure sea water and stucco for wet climates. Different kinds of modern cement, most of them known as Portland cement or blends, including blast furnace cement, Portland fly-ash cement, Portland pozzolan cement, pozzolan-lime cement, slag-lime cement etc.

Cement Chemistry

Cement is chiefly of two kinds based on the way it is set and hardened: hydraulic cement, which hardens due to the addition of water, and non-hydraulic cement, which is hardened by carbonation with the carbon present in the air, so it cannot be used underwater.

Non-hydraulic cement is produced through the following steps (lime cycle):

1. Calcination: Lime is produced from limestone at over 825° C for about 10 hours. ($CaCO_3 \rightarrow CaO + CO_2$)

2. Slaking: Calcium oxide is mixed with water to make slaked lime. ($CaO + H_2O \rightarrow Ca(OH)_2$)

3. Setting: Water is completely evaporated.

4. The cement is exposed to dry air and it hardens after time-consuming reactions. ($Ca(OH)_2 + CO_2 \rightarrow CaCO_3 + H_2O$)

On the other hand, hydraulic cement is mainly made up of silicates and oxides:

1 Belite ($2CaO \cdot SiO_2$);

2 Alite ($3CaO \cdot SiO_2$);

3 Tricalcium aluminate/ Celite ($3CaO \cdot Al_2O_3$)

4 Brownmillerite ($4CaO \cdot Al_2O3 \cdot Fe_2O_3$)

The ingredients are processed in the kiln in cement plants. Complete chemistry of the reactions is still a subject of research.

To Prepare Cement Mortar

Cement mortar is like a paste which is prepared by adding certain quantity of water to cement and sand mixture. Cement in this case is denoted as matrix while sand is termed as adulterant.

We know Cement has good binding properties while there are other binding materials are also available, but cement is mostly used because of its high strength and water resisting properties. It is used to create a strong bond between bricks, stones in a masonry.

Plastering is done by cement mortar which gives smooth finish to the structure. Cement molds of different shapes can be made using cement mortar. It is also used to seal the joints of brickwork and stone work or cracks.

Generally, the cement sand ratio in a mortar is in between 1:2 to 1:6. The ratio of cement and sand mix is decided based on the importance of work.

Uses of Cement

To Prepare Cement Concrete

Cement concrete is a major building material in the world which is widely using because of its marvelous structural properties. The ingredients of cement concrete are cement, fine aggregate, coarse aggregate and water respectively.

In general, ordinary Portland cement is used to prepare concrete. But for special cases or based on different circumstances many types of cements like rapid hardening cement, high alumina cement etc. are discovered.

To Build Fire Proof or Heat Proof Structures

To with stand against high temperatures and to prevent fire accidents structures should be built with great fire-resistant materials like cement. High alumina cement is more suitable material to make concrete for the structures in high temperature regions.

To Build Hydrographic and Frost Resistant Structures

Most of the hydrographic structures in the world are built using concrete with cement as binding material. The structures built in water or in contact with water should be very strong against moisture and they should be water tight.

Many types of cements like hydrophobic cement, expanding cement, pozzolana cement, quick setting cement etc. are most suitable for constructing water-retaining structures. Quick setting cement is very much useful in the case when there is limited time to construct under water structures.

Hydrophobic cement had more resistance against frost actions, so it can be used to build structures in snow regions also.

To Build Chemical Proof Structures

In chemical industries, different chemicals are stored and they may damage the structure if proper resistance is not there. Acid resistant cement is very much useful in this case.

Similarly, for the constructions under marine conditions, sewage carrying structures etc. Sulphate resistant cement is useful.

For Grouting

Grouting is the process of filling cracks, joints, openings in foundations or any other structural members to improve their strength. In general, ordinary Portland cements is used as grout material to which required amount of water and sand is added.

To fill very fine cracks or to fill deep thin cracks micro fine cement is most suitable. Micro fine cement contains very finer particles than the ordinary Portland cement so it can flow into very fine and deep cracks in quick time.

To Construct Cement Concrete Roads

Cement concrete roads are more famous as high standard roads which are stronger than all other types of roads. They are also called as rigid pavements because of their rigid nature. C.C roads have long life span even without proper maintenance. Load wise also they are much capable than all other types.

To Manufacture Precast Members

Many precast members are made using cement as binding material. Cement concrete pipes are widely used as drains, pipes under culverts etc. Cement concrete brick masonry is more famous because of the size of block, ease of construction and strength etc.

Water tanks and septic tanks are generally constructed by cement concrete rings. Many other things like garden seats, flower pots, dust bins, lam posts etc. are manufactured using cement.

For Aesthetic Structures

Now days cement is available in many colors. This is done by adding coloring agent while manufacturing cement but the percentage of coloring agent should be below 10%. Some of the coloring agents are iron oxide which gives red or brown, cobalt which gives blue etc.

The colored cement makes the structure beautiful without any painting. Colored cements generally used for floor finishing, stair treads, window sill slabs, external wall surfaces etc.

Portland Cement

Portland cement is a basic cement mixture and a fundamental ingredient for many common cement applications. Because Portland cement is very common, it is also often called "Ordinary Portland Cement" or simply "OPC." Since the early 20th century, Portland cement has been used internationally for a wide variety of applications including concrete projects, mortar pastes, stucco decorations, and grout fillings.

"Cement" and "concrete" often have synonymous meaning in casual conversations, strictly defined, however, they mean very different things. Concrete contains cement, but cement does not contain concrete. Cement is the binding agent used to create concretes, mortars, and grouts. Alternatively, rocks, gravels, sands, and water are combined with cement to create the hard, stony masses known as concrete. Portland cement concrete is concrete made with Portland cement as the binding agent.

Modern Portland cement does not mean cement that comes from Portland. Although the original Portland cement was made using stones extracted from Britain's Portland Isle, the modern use of the term Portland cement is much more general. Today, Portland cement is a term for a standard, generic binding cement.

Think of Portland cement simply as ordinary cement you see used in driveways, commercial buildings, and homes every day.

Roman cement was the predecessor to ordinary Portland cement. A crude form of Roman cement was rediscovered by English engineer John Smeaton in 1756. The formula for Roman cement was refined and patented in 1796 by James Parker.

Around 1811, James Frost developed British cement: a new way of processing cement that became instrumental in making Portland cement. In 1824, the first form of Portland cement was patented by English cement manufacturer Joseph Aspdin. Aspdin's version of Portland cement was further refined by his son, William, in 1843, and it closely resembled the ordinary Portland cement we know today.

Isaac Charles Johnson was another English cement manufacturer who competed with William Aspdin for the title of "true Portland cement creator." However, the exact history revealing who developed it first remains unclear.

The United States imported Portland cement from Germany and England as early as 1868. Between the 1970s and 1980s, the United States began manufacturing their Portland cement in Michigan and Pennsylvania. Within a few years, most of the Portland cement used in the United States was produced locally.

The Ingredients in Portland Cement

What constitutes Portland cement slightly varies depending on the organization that defines it. It is a hydraulic lime cement, meaning water is added to a clinker mixture which solidifies into a water-resistant material. Portland clinker is a combination of raw materials heat treated then ground or pulverized into cement powders. The most common materials used in Portland clinker are limestone, clays, and sands.

Although most of a Portland cement mixture is clinker aggregate (usually 90% by mass), there are also tiny amounts of various admixtures that help control factors like curing time, for example. Calcium sulfate (gypsum) and magnesium oxide are (magnesia) are two examples of admixtures often found in Portland cement mixtures. Other common basic components of Portland cements are alumina, iron, calcium and silica.

Types of Portland Cements

Type 1: The most common type of Portland cement for general applications. Type 1 Portland cement is the standard cement used for many precast and pre-stressed construction projects (for example, buildings, bridges, pavements, and much more). This type has a high mass of Tricalcium silicate and is not recommended for use with projects that rest on soil or in groundwater.

Type 2: Another general construction concrete, but moderate resistance to soil and groundwater contact. Because type 2 Portland cement has less Tricalcium aluminate than type 1 cement, it can tolerate more direct contact with soils containing higher levels of sulfate ions, such as those on the Pacific coast in America.

Type 3: The best for cold-weather construction applications. Type 3 Portland Cement sets more quickly than types 1 or 2 because it requires less hydration. This feature is ideal for cold climates because wintry weather reduces atmospheric hydration. Furthermore, the early high-strength of type 3 cement allows for more immediate use after construction projects.

Type 4: Used for large/thick concrete structures. Type 4 Portland cement cures slower than types 1-3, therefore generally doesn't reach maximum strength for a couple of years. However once set, type 4 Portland cement is stronger than types 1-3. Dams once used type 4 Portland cement, for example, but innovations in cement manufacturing now offer better alternatives to type 4 Portland cement.

Type 5: The best sulfate resistance for construction that contacts alkali soil and ground water sulfates. Although type 5 Portland cement is still used around areas near the Pacific, they are becoming less popular because admixtures can be used with ordinary cements with comparable results.

Geopolymer Cement

Geopolymer cement is a binding system that hardens at room temperature, like regular Portland cement. If a geopolymer compound requires heat setting it may not be called geopolymer cement but rather geopolymer binder.

Geopolymer cement is an innovative material and a real alternative to conventional Portland cement for use in transportation infrastructure, construction and offshore applications. It relies on minimally processed natural materials or industrial byproducts to significantly reduce its carbon footprint, while also being very resistant to many of the durability issues that can plague conventional concretes.

Creating geopolymer cement requires an alumina silicate material, a user-friendly alkaline reagent (sodium or potassium soluble silicates with a molar ratio MR $SiO_2:M_2O>1,65$, M being Na or K) and water. Room temperature hardening relies on the addition of calcium cations, essentially iron blast furnace slag.

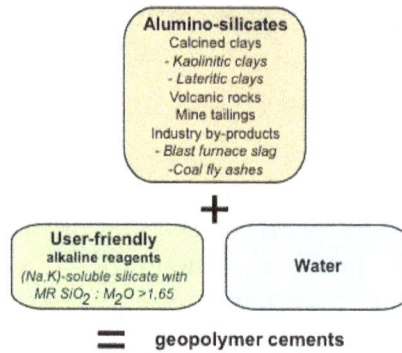

Geopolymer cements cure more rapidly than Portland-based cements. They gain most of their strength within 24 hours.

However, they set slowly enough that they can be mixed at a batch plant and delivered in a concrete mixer. Geopolymer cement also has the ability to form a strong chemical bond with all kind of rock-based aggregates. On March 2010, the US Department of Transportation Federal Highway Administration released a TechBrief titled Geopolymer Concrete that states: The production of versatile, cost-effective geopolymer cements that can be mixed and hardened essentially like Portland cement represents a game changing advancement, revolutionizing the construction of transportation infrastructure and the building industry.

Geopolymer Concrete

There is often confusion between the meanings of the two terms 'geopolymer cement' and 'geopolymer concrete'. A cement is a binder whereas concrete is the composite material resulting from the addition of cement to stone aggregates. In other words, to produce concrete one purchases cement (generally Portland cement or Geopolymer cement) and adds it to the concrete batch. Geopolymer chemistry was from the start aimed at manufacturing binders and cements for various types of applications. For example the British company banah UK sells its *banah-Cem*™ as geopolymer cement, whereas the Australian company Zeobond markets its *E-crete*™ as geopolymer concrete.

Portland Cement Chemistry vs Geopolymer Cement Chemistry

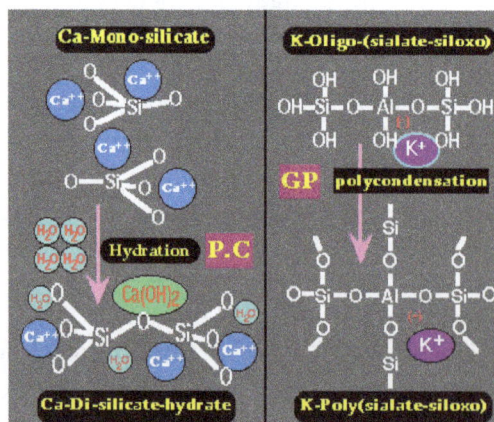

Left: hardening of Portland cement (P.C.) through simple hydration of Calcium Silicate into Calcium Di-Silicate hydrate (CSH) and lime $Ca(OH)_2$.

Right: hardening (setting) of Geopolymer cement (GP) through poly- condensation of Potassium Oligo- (sialate-siloxo) into Potassium Poly(sialate-siloxo) cross linked network.

Alkali-activated Materials vs Geopolymer Cements

Geopolymerization chemistry requires appropriate terminologies and notions that are evidently different from those in use by Portland cement experts. "Joseph Davidovits developed the notion of a geopolymer (a Si/Al inorganic polymer) to better explain these chemical processes and the resultant material properties. To do so required a major shift in perspective, away from the classical crystalline hydration chemistry of conventional cement chemistry. To date this shift has not been well accepted by practitioners in the field of alkali activated cements who still tend to explain such reaction chemistry in Portland cement terminology.

Indeed, geopolymer cement is sometimes mixed up with alkali-activated cement and concrete, developed more than 50 years ago by G.V. Glukhovsky in Ukraine, the former Soviet Union. They were originally known under the names "soil silicate concretes" and "soil cements". Because Portland cement concretes can be affected by the deleterious *Alkali-aggregate reaction*, coined AAR or Alkali-silica reaction coined ASR, the wording *alkali-activation* has a negative impact on civil engineers. Nevertheless, several cement scientists continue to promote the idea of *alkali-activated materials* or *alkali-activated geopolymers*. These cements coined AAM encompass the specific fields of alkali-activated slags, alkali-activatedcoal fly ashes, blended cements. However, it is interesting to mention the fact that geopolymer cements do not generate any of these deleterious reactions.

User-friendly Alkaline-reagents

Although geopolymerization does not rely on toxic organic solvents but only on water, it needs chemical ingredients that may be dangerous and therefore requires some safety procedures. Material Safety rules classify the alkaline products in two categories: corrosive products and irritant products. The two classes are recognizable through their respective logos. The corrosive products must be handled with gloves, glasses and masks. They are User- hostile and cannot be implemented in mass applications without the appropriate safety procedures. In the second category one finds Portland cement or hydrated lime, typical mass products. Geopolymeric alkaline reagents belonging to this class may also be termed as User-friendly.

Unfortunately, the development of so-called alkali-activated-cements or alkali-activated geopolymers (the latter being a wrong terminology), as well as several recipes found in the literature and on the Internet, especially those based on fly ashes, comprise molar ratio below 1.20, in average below 1.0. Worse, looking only at low-costs consideration, not at safety and User- friendly issues, they propose systems based on pure NaOH (8M or 12M). These are User- hostile conditions and may not be used by the ordinary labor force employed in the field. Indeed, laws, regulations, and state directives push to enforce for more health protections and security protocols for workers' safety.

🙁 hostile	friendly 🙂
CaO (quick lime) NaOH, KOH	Ca(OH)2 Portland cement, Iron slag
Sodium metasilicate $SiO_2:Na_2O = 1.0$	Slurry soluble silicate/kaolin $1.25 < SiO_2:Na_2O < 1.45$
Any soluble silicate $SiO_2:Na_2O < 1.45$	Any soluble silicate $SiO_2:Na_2O > 1.45$

On the opposite, Geopolymer cement recipes employed in the field generally involve alkaline soluble silicates with starting molar ratio $SiO_2:M_2O$ ranging from 1.45 to 1.95, essentially 1.45 to 1.85, i.e. user-friendly conditions. It may happen that for research, some laboratory recipes have molar ratios in the 1.20 to 1.45 range. Yet, this is only for study, not for manufacture.

Geopolymer Cement Categories

The categories comprise:

- Slag-based geopolymer cement.

- Rock-based geopolymer cement.

- Fly ash-based geopolymer cement

  o type 1: alkali-activated fly ash geopolymer.

  o type 2: slag/fly ash-based geopolymer cement.

- Ferro-sialate-based geopolymer cement.

Slag-based Geopolymer Cement

Manufacture components: metakaolin MK-750 + blast furnace slag + alkali silicate (user-friendly).

Geopolymeric make-up: Si:Al = 2 in fact solid solution of Si:Al=1, Ca-poly(di- sialate) (anorthite type) + Si:Al =3 , K-poly(sialate-disiloxo) (orthoclase type) and CSH Ca-disilicate hydrate.

The first geopolymer cement developed in the 1980s was of the type (K,Na,Ca)- poly(sialate) (or slag-based geopolymer cement) and resulted from the research developments carried out by J. Davidovits and J.L. Sawyer at Lone Star Industries, USA and yielded to the invention of the well known Pyrament cement. The American patent application was filed in 1984 and the patent US 4,509,985 was granted on April 9, 1985 with the title 'Early high-strength mineral polymer'.

Rock-based Geopolymer Cement

The replacement of a certain amount of MK-750 with selected volcanic tuffs yields geopolymer cement with better property and less CO2 emission than the simple slag- based geopolymer cement.

Manufacture components: metakaolin MK-750, blast furnace slag, volcanic tuffs (calcined or not calcined), mine tailings and alkali silicate (user-friendly).

Geopolymeric make-up: Si:Al = 3, in fact solid solution of Si:Al=1 Ca-poly(di- sialate) (anorthite type) + Si:Al =3-5 (Na,K)-poly(silate-multisiloxo) and CSH Ca- disilicate hydrate.

Fly ash-based Geopolymer Cements

Later on, in 1997, building on the works conducted on slag-based geopolymeric cements, on the one hand and on the synthesis of zeolites from fly ashes on the other hand, Silverstrim and van Jaarsveld and van Deventer developed geopolymeric fly ash-based cements. Silverstrim et al. US Patent 5,601,643 was titled 'Fly ash cementitious material and method of making a product'.

Presently two types based on Class F fly ashes:

- Type 1: alkali-activated fly ash geopolymer (user-hostile):

  In general requires heat hardening at 60-80°C and is not manufactured separately and becomes part of the resulting fly-ash based concrete. NaOH (user-hostile) + fly ash: fly ash particles embedded in an alumino-silicate gel with Si:Al= 1 to 2, zeolitic type (chabazite-Na and sodalite).

- Type 2: slag/fly ash-based geopolymer cement (user-friendly):

  Room-temperature cement hardening. User-friendly silicate solution + blast furnace slag + fly ash: fly ash particles embedded in a geopolymeric matrix with Si:Al= 2, (Ca,K)-poly(sialate-siloxo).

Ferro-sialate-based Geopolymer Cement

The properties are similar to those of Rock-based geopolymer cement but involve geological elements with high iron oxide content. The geopolymeric make up is of the type poly(ferro-sialate) (Ca,K)-(-Fe-O)-(Si-O-Al-O-). This user-friendly geopolymer cement is on the development and pre-industrialization phase.

CO_2 Emissions During Manufacture

Concrete (mixture of cement and aggregates) is the most commonly used construction material; its usage by communities across the globe is second only to water. Ever grander building and infrastructure projects require prodigious quantities of concrete with its binder of Portland cement whose manufacture is accompanied by large emissions of carbon dioxide CO_2. According to the Australian concrete expert B. V. J. Rangan, this burgeoning worldwide demand for concrete is a great opportunity for the development of geopolymer cements of all types, with their much lower tally of carbon dioxide CO_2.

CO_2 Emission During Manufacture of Portland Cement Clinker

Ordinary cement, often called by its formal name of Portland cement, is a serious atmospheric pollutant. Indeed, the manufacture of Portland cement clinker involves the calcination of calcium carbonate according to the reaction:

$$5CaCO_3 + 2SiO_2 \rightarrow (3CaO,SiO_2)(2CaO,SiO_2) + 5CO$$

The production of 1 tonne of Portland clinker directly generates 0.55 tonnes of chemical-CO2 and requires the combustion of carbon-fuel to yield an additional 0.40 tonnes of carbon dioxide.

To simplify: 1 *T of Portland cement = 0.95 T of carbon dioxide*

The only exceptions are so-called 'blended cements', using such ingredients as coal fly ash, where the CO_2 emissions are slightly suppressed, by a maximum of 10%- 15%. There is no known technology to reduce carbon dioxide emissions of Portland cement any further.

On the opposite, Geopolymer cements do not rely on calcium carbonate and generate much less CO_2 during manufacture, i.e. a reduction in the range of 40% to 80-90%.

The Portland cement industry reacted strongly by lobbying the legal institutions so that they delivered CO_2 emission numbers, which did not include the part related to calcium carbonate decomposition, focusing only on combustion emission. estimates for CO_2 emissions from cement production have concentrated only on the former source. The UN's Intergovernmental Panel on Climate Change puts the industry's total contribution to CO_2 emissions at 2.4%; the Carbon Dioxide Information Analysis Center at the Oak Ridge National Laboratory in Tennessee quotes 2.6%. Now Joseph Davidovits of the Geopolymer Institute has for the first time looked at both sources. He has calculated that world cement production of 1.4 billion tonnes a year produces 7% of current CO_2 emissions. Fifteen years later, the situation has worsened with Portland cement CO_2 emissions approaching 3 billion tonnes a year.

The fact that the dangers to the world's ecological system from the manufacture of Portland cement is so little known by politicians and public makes the problem all the more urgent: when nothing is known, nothing is done. This situation clearly cannot continue if the world is going to survive.

Geopolymer Cements Energy Needs and CO_2 1issions

There is a comparison between the energy needs and CO_2 emissions for regular Portland cement, Rock-based Geopolymer Cements and Fly ash-based geopolymer cements. The comparison proceeds between Portland cement and geopolymer cements with similar strength, i.e. average 40 MPa at 28 days. There have been several studies published on the subject that may be summarized in the following way:

Rock-based Geopolymer cement manufacture involves:

- 70% by weight geological compounds (calcined at 700° C)
- blast furnace slag

- alkali-silicate solution (industrial chemical, user-friendly).

The presence of blast furnace slag provides room-temperature hardening and increases the mechanical strength.

Energy needs and CO_2 emissions for 1 tonne of Portland cement and Rock-based Geopolymer cement.

Energy needs (MJ/tonne)	Calcination	Crushing	Silicate Sol.	Total	Reduction
Portland Cement	4270	430	0	4700	0
GP-cement, slag by-product	1200	390	375	1965	59%
GP-cement, slag manufacture	1950	390	375	2715	43%
CO2 emissions (tonne)					
Portland Cement	1.000	0.020		1.020	0
GP-cement, slag by-product	0.140	0,018	0.050	0.208	80%
GP-cement, slag manufacture	0.240	0.018	0.050	0.308	70%

Energy needs

According to the US Portland Cement Association , energy needs for Portland cement is in the range of 4700 MJ/tonne (average). The calculation for Rock- based geopolymer cement is performed with following parameters:

- the blast furnace slag is available as by-product from the steel industry (no additional energy needed);

- or must be manufactured (re-smelting from non granulated slag or from geological resources).

In the most favorable case — slag availability as by-product — there is a reduction of 59% of the energy needs in the manufacture of Rock-based geopolymer- cement in comparison with Portland cement.

In the least favorable case —slag manufacture — the reduction reaches 43%.

CO_2 emissions during manufacture

In the most favorable case — slag availability as by-product — there is a reduction of 80% of the CO_2 emission during manufacture of Rock-based geopolymer cement in comparison with Portland cement.

In the least favorable case —slag manufacture — the reduction reaches 70%.

Fly ash-based Cements Class F Fly Ashes

They do not require any further heat treatment. The calculation is therefore easier. One achieves emissions in the range of 0,09 to 0,25 tonnes of CO_2 / 1 tonne of fly ash-based cement, i.e. CO_2 emissions that are reduced in the range of 75 to 90%.

Properties for Rock-based Geopolymer Cement
(Ca,K)- poly(Sialate-disiloxo)

- Shrinkage during setting: < 0.05%, not measurable.

- Compressive strength (uniaxial): > 90 MPa at 28 days (for high early strength formulation, 20 MPa after 4 hours).

- Flexural strength: 10–15 MPa at 28 days (for high early strength 10 MPa after 24 hours).

- Young modulus: > 2 GPa.

- Freeze-thaw: mass loss < 0.1% (ASTM 4842), strength loss <5% after 180 cycles.

- Wet-dry: mass loss < 0.1% (ASTM 4843).

- Leaching in water, after 180 days: k20 < 0.015%.

- Water absorption: < 3%, not related to permeability.

- Hydraulic permeability: 10^{-10} m/s.

- Sulfuric acid, 10%: mass loss 0.1% per day.

- Chlorhydric acid 5%: mass loss 1% per day.

- KOH 50%: mass loss 0.02% per day.

- Ammonia solution: no mass loss.

- Sulfate solution: shrinkage 0.02% at 28 days.

- Alkali-aggregate reaction: no expansion after 250 days (-0.01%), as shown in the graph, comparison with Portland cement (ASTM c227). These results were published as earlier as 1993. Geopolymer binders and cements even with alkali contents as high as 10%, do not generate any dangerous alkali-aggregate reaction.

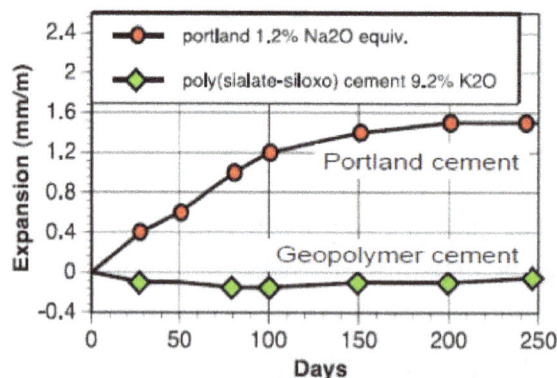

The innocuity towards Alkali-Aggregate Reaction is always confirmed in geopolymer cements. More recently Li et al. used another standard, ASTM C 441-97, by which powdered quartz glass is the reactive fine element. The test duration is 90 days. Portland cement mortars exhibited expansion at 90 days in the range of 0.9– 1.0 % whereas geopolymer cement remained practically unchanged, with a small shrinkage of -0.03 % at 90 days.

The Need for Standards

In June 2012, the institution ASTM International (former American Society for Testing and Materials, ASTM) organized a symposium on Geopolymer Binder Systems. The introduction to the symposium states: When performance specifications for Portland cement were written, non-portland binders were uncommon. New binders such as geopolymers are being increasingly researched, marketed as specialty products, and explored for use in structural concrete. This symposium is intended to provide an opportunity for ASTM to consider whether the existing cement standards provide, on the one hand, an effective framework for further exploration of geopolymer binders and, on the other hand, reliable protection for users of these materials.

The existing Portland cement standards are not adapted to geopolymer cements. They must be created by an ad hoc committee. Yet, to do so, requires also the presence of standard geopolymer cements. Presently, every expert is presenting his own recipe based on local raw materials (wastes, by-products or extracted). There is a need for selecting the right geopolymer cement category. The 2012 State of the Geopolymer R&D, suggested to select two categories, namely:

- Type 2 slag/fly ash-based geopolymer cement: fly ashes are available in the major emerging countries;

and

- Ferro-sialate-based geopolymer cement: this geological iron rich raw material is present in all countries through out the globe.

and

- the appropriate user-friendly geopolymeric reagent.

Fibre Cement

Fibre cement has been around for over 100 years but the technology to create better fibre cement building products has been improving since that time, especially in the wake of massive change in building regulations regarding the use of asbestos in recent times.

Fibre cement composites are composed of Portland cement, silica, water and wood pulp and are manufactured using the so-called " Hatschek process". This process was initially developed for the production of asbestos composites, but is now used for something that can actually be used as a viable asbestos replacement material: fibre reinforced cement composites, or fibre cement for short.

As part of the Hatscheck process, cellulose fibers are pulped in warm water then mixed with cement, silica, and other additives. The fibre cement mixture is transferred onto a conveyor belt, then a laminating roll. This process is repeated and multiple layers of fibre cement are laminated to the required thickness.

The laminated FC sheet is cured using an autoclave process which combines intense heat and pressure to enhance chemical reaction between cement and silica to form fibre cement composites, which are extremely tough building materials. This curing process results in fibre cement composites with superior properties compared to air-cured fiber cement composites, for example higher strength and toughness, low moisture movement, low alkalinity, high fire resistance and good workability.

Pigment can also be added to make fibre cement products of various colours. Fibre cement building products have many benefits over traditional materials such as wooden planks or gypsum or plywood boards.

Fibre cement boards and planks are highly durable, flexible, water resistant, fire proof, resistant to insects and chemical corrosion. They are also highly workable using normal tools; in short: they are useful in all kinds of buildings and throughout the whole house, from the cellar to the roof.

Typical applications of fibre cement building materials include use as substrates, internal/external cladding and use in wet areas where water resistance is required, as well as areas where weather and fire resistance is important.

Rosendale Cement

Rosendale cement was the most famous cement of its day and was the benchmark for all other natural cements. Between 1870 and 1872, Rosendale cement was used in building the piers for the Brooklyn Bridge. These massive 45 and 78-feet deep foundations were built of solid concrete. The 276-feet masonry towers also used mortar made of Rosendale cement. The towers took three more years to complete and were higher than New York's tallest office building at the time.

From 1884 to 1886 Rosendale cement from the Widow Jane mine was carried on the D&H Canal thence down the Hudson River to Liberty Island where it was used to construct the base for the Statue of Liberty. This was the largest 19th century concrete structure in the United States. The pedestal foundation extends to a depth of 20 feet and is almost all solid concrete. Massive concrete walls 8 to 19 feet thick back the granite walls of the pedestal.

D&H Canal coal was important to the 19th century construction industry. It fueled the Rosendale cement kilns and was a key ingredient in the clay mix used by the Hudson Valley brickyards. These two industries complimented one another. Rosendale cement was an excellent material for the mortar used with Hudson River bricks and didn't compete with brick construction because it wasn't good for large concrete castings. It had a slow chemical reaction, set slowly and was prone to shrinkage cracks in large castings because of its evaporative drying.

In 1871, Portland cement technology finally came to the United States. Limestone deposits discovered in the Lehigh Valley of eastern Pennsylvania proved conducive to manufacturing Portland cement. By 1900, the Hudson Valley also had two operating Portland plants. Portland cement set very rapidly and was soon preferred over Rosendale cement for mortar. Cast concrete also didn't have time to develop shrinkage cracks and, when used with steel framing, became a viable substitute for brick and mortar construction.

Portland cement had a significant impact on the Rosendale cement market. From 1900 to 1910, natural cement production across the United States plummeted from about 10 million barrels annually to one million while Portland cement production grew from one million barrels in 1895 to over 12 million by 1910.

The D&H Canal was critical in establishing and sustaining the Rosendale cement industry. In turn, New York City was the impetus behind building the canal in the first place. The city's growth required that it reach out to its hinterland for a continuing supply of building materials. During the 19th century a steady flow of wood, bricks, cement and bluestone was carried down the Hudson to the city alongside the coal that would be used to heat the growing metropolis. In a sense there was a pipeline from the sources of New York's key commodities to the city itself. The city helped foster these industries and provided a market for them.

The industries the canal helped spawn eventually came to overshadow the canal itself. The Rosendale cement industry is a prime example. When the canal was finally shut down in 1898 the section from Rosendale to the canal's terminus at Eddyville was kept open until 1913 in order to carry cement to the Hudson River boats. In this regard it was the cement industry that in the end supported the canal and not vice versa.

Soil Cement

Soil-cement is a highly compacted mixture of soil/aggregate, cement, and water. It is widely used as a low-cost pavement base for roads, residential streets, parking areas, airports, shoulders, and materials-handling and storage areas. Its advantages of great strength and durability combine with low first cost to make it the outstanding value in its field. A thin bituminous surface is usually placed on the soil-cement to complete the pavement.

Soil-cement is sometimes called cement-stabilized base, or cement-treated aggregate base. Regardless of the name, the principles governing its composition and construction are the same.

Type of Soil Used

The soil material in soil-cement can be almost any combination of sand, silt, clay, gravel, or crushed stone. Local granular materials, such as slag, caliche, limerock, and scoria, plus a wide variety of waste materials including cinders, fly ash, foundry sands, and screenings from quarries and gravel pits, can all be utilized as soil material. Old granular-base roads, with or without bituminous surfaces, can also be reclaimed to make great soil-cement.

Preparation of Soil Cement

Before construction begins, simple laboratory tests establish the cement content, compaction, and water requirements of the soil material to be used. During construction, tests are made to see that the requirements are being met. Testing ensures that the mixture will have strength and long-term durability.

Soil-cement can be mixed in place or in a central mixing plant. Central mixing plants can be used where borrow material is involved. Friable granular materials are selected for their low cement requirements and ease of handling and mixing. Normally pugmill-type mixers are used. The mixed soil-cement is then hauled to the jobsite and spread on the prepared subgrade.

Compaction and curing procedures are the same for central-plant and mixed-in-place procedures.

There are four steps in mixed-in-place soil-cement construction; spreading cement, mixing, compaction, and curing. The proper quantity of cement is spread on the in-place soil material. Then the cement, the soil material, and the necessary amount of water are mixed thoroughly by any of several types of mixing machines. Next, the mixture is tightly compacted to obtain maximum benefit form the cement. No special compaction equipment is needed; rollers of various kinds, depending on soil type, can be used. The mixture is cemented permanently at a high density and the hardened soil-cement will not deform or consolidate further under traffic.

Curing, the final step, prevents evaporation of water to ensure maximum strength development through cement hydration. A light coat of bituminous material is commonly used to prevent moisture loss; it also forms part of the bituminous surface. A common type of wearing surface for light traffic is a surface treatment of bituminous material and chips 0.5- to 0.75-inch thick. For heavy-duty use and in severe climates a 1.5-inch asphalt mat is used.

Contractors bidding on soil-cement jobs know that construction will be relatively easy and problem-free; weather delays rare; and reworking of completed sections unnecessary.

Advantages of Soil-cement

Failing granular-base pavements, with or without their old bituminous mats, can be salvaged, strengthened, and reclaimed as soil-cement pavements. This is an efficient, economical way of rebuilding pavements. Since approximately 90 percent of the material used is already in place, handling and hauling costs are cut to a minimum. Many granular and waste materials from quarries and gravel pits can also be used to make soil-cement; thus high-grade materials are conserved for other purposes.

Highway and city engineers praise soil-cement's performance, its low first cost, long life, and high strength. Soil-cement is constructed quickly and easily – a fact appreciated by owners and users alike.

Performance

Soil-cement thicknesses are less than those required for granular bases carrying the same traffic over the same subgrade. This is because soil-cement is a cemented, rigid material that distributes loads over broad areas. Its slab-like characteristics and beam strength are unmatched by granular bases. Hard, rigid soil-cement resists cyclic cold, rain, and spring-thaw damage.

Old soil-cement pavements in all parts of the continent are still giving good service at low maintenance costs. Soil-cement has been used in every state in the United States and in all Canadian provinces. Specimens taken from roads show that the strength of soil-cement actually increases with age; some specimens were four times as strong as test specimens made when the roads were first opened to traffic. This reserve strength accounts in part for soil-cement's good long-term performance.

Cost

The cost of soil-cement compares favorably with that of granular-base pavement. When built for equal load-carrying capacity, soil-cement is almost always less expensive than other low-cost pavements. Economy is achieved through the use or reuse of in-place or nearby borrow materials. No costly hauling of expensive, granular-base materials is required; thus, both energy and materials are conserved.

Pavement Applications

Soil-cement pavements have many uses from city streets, county roads, state routes, and interstate highways, to parking lots, industrial storage facilities, and airports. In fact, the "family" of soil-cement pavement products can actually be divided up into three main components – each with their own unique contribution to a pavement structure. These components include Cement-Modified Soils (CMS), Cement-Treated Base (CTB), and Full-Depth Reclamation (FDR).

White Portland Cement

White Portland cement, as the name indicates, is a kind of cement with white color. It is the same as ordinary gray Portland cement except in respect to color and fineness.

Snowcrete is the commercial name given to White Cement.

The color of a structure is very important in the perspective of the architectural point of view. White cement produces a concrete with perfect and uniform color throughout. It is possible to produce very light shades of pastels and other colors by adding pigments with white cement which are not possible with normally used Gray Portland Cement. Besides, there are various uses of White Portland Cement.

Other main features of White Portland Cement are given below:

- The color of cement is an especially important issue in the white cement. Therefore, the principal concerns are consistency in brightness and tone.

- The quality of raw materials and the manufacturing process affects the color of the white cement.

- This cement is made from raw materials containing very little Iron Oxide and Manganese Oxide. These oxides influence whiteness and undertone of white Portland cement.

- Generally in white cement China clay is used, together with chalk or limestone.

- The oil used in white cement as fuel for fuel for the kiln in order to avoid contamination by coal ash.

- The manufacturing process of white cement should be controlled with special precautions.

- White cement's chemical composition and physical characteristics meet the specifications of Type I Portland cement.

References

- Silberstein, Eugene (2004). Residential construction academy: HVAC. Residential Construction Academy Series. Cengage Learning. p. 467. ISBN 978-1-4018-4901-6. Retrieved 2009-09-21

- Important-properties-of-concrete: civilengineerspk.com, Retrieved 13 June 2018

- What-is-precast-concrete: nitterhouseconcrete.com, Retrieved 31 March 2018

- Fiber-reinforced-concrete-150: theconstructor.org, Retrieved 21 March 2018

- Grady, Joe (2004-06-01). "The finer points of bonding to gypsum concrete underlayment". National Floor Trends. Retrieved 2009-09-21

- Benefits-of-polished-concrete, polished-concrete: ardorsolutions.com, Retrieved 10 July 2018

- What-is-portland-cement: merloconstructionmi.com, Retrieved 20 June 2018

- The-rosendale-cement-industry: neversinkmuseum.org, Retrieved 16 May 2018

3

Steel and Wood

Steel is an important construction and engineering material. It is an alloy of the elements iron, carbon, manganese, nickel, chromium and other elements in trace amounts. Wood is another common building material used for the construction of roofs, interior doors and frames, floor, etc. The topics elaborated in this chapter on structural steel, I-beam, steel channel, lumber, timber framing, etc. will help in providing an understanding of the use of steel and wood in construction.

Structural Steel

Structural steel is a standard construction material made from specific steel grades and is available in industry standard cross sectional shapes. This steel exhibits desirable physical properties such as strength, uniformity of properties, light weight and ease of use etc. This makes it one of the most versatile structural materials in use. Major applications for these steels are in high-rise and tall multi-storey buildings, industrial buildings, towers, tunnels, bridges, road barriers and industrial structures etc.

The Various Types of Structural Steel Shapes

American Standard Beam (S-Shaped)

Generally known as an S beam, the American standard beam has a rolled section with two parallel flanges, all connected by a web. The flanges on S-shaped beams are relatively narrow. The designation of the beam gives the builder information about each unit's width and weight. For example, S12x50 represents a beam that's 12 inches deep and weighs 50 pounds per foot.

Angle (L-Shaped)

Angle beams take an L shape, with two legs that come together at a 90-degree angle. Angle beams come in equal or unequal leg sizes. An unequal leg L beam may have one leg of 2x2x0.5 and one leg of 6x3x0.5, for example. L beams are typically used in floor systems because of the reduced structural depth.

Bearing Pile (H-Shaped)

When builders can't find a structure on a shallow foundation, they use bearing piles to design a deep foundation system. Bearing piles are H-shaped to effectively transfer loads through the pile to the tip. Bearing piles work best in dense soils that offer most resistance at the tip. Individual piles can bear more than 1,000 tons of weight.

Channel (C-Shaped)

Structural C channels, or C beams, have a C-shaped cross section. Channels have top and bottom flanges, with a web connecting them. C-shaped beams are cost-effective solutions for short- to medium-span structures. Channel beams were originally designed for bridges, but are popular for use in marine piers and other building applications.

Hollow Steel Section (HSS)

HSS is a metal profile that has a hollow, tubular cross section. HSS units can be square, rectangular, circular, or elliptical. HSS structures are rounded, with radiuses that are about twice the thickness of the wall. Engineers commonly use HSS sections in welded steel frames for which units experience loading in different directions.

I-Beam

An I Beam, also known as an H beam or a universal beam, has two horizontal elements, the flanges, with a vertical element as the web. The web is capable of resisting shear forces, while the horizontal flanges resist most of the beam's bending movement. The I shape is very effective at carrying shear and bending loads in the web's plane. The construction industry widely uses I beams in a variety of sizes.

Pipe

Structural steel pipes are important for a variety of construction applications, lending strength and stability. Pipes are hollow, cylindrical tubes that come in a variety of sizes. Engineers often use steel pipes to meet the needs of water, oil, and gas industry projects.

Tee

A tee beam, or T beam, is a load-bearing beam with a T-shaped cross section. The top of this cross section is the flange, with the vertical web below. Tee beams can withstand large loads but lack the bottom flange of the I Beam, giving it a disadvantage in some applications.

Custom Shapes

Today's engineers are not limited to using only the most common shapes. Custom metal fabrication opens the doors to a variety of special structural steel shapes for any type of project. Using state-of-the-art tools and techniques, such as water jet, laser, and plasma cutting, metal fabricators can sculpt steel into myriad shapes for specific needs.

Types of Structural Steel

Structural grade steels have specific chemical compositions and mechanical properties required as per their application. These steels are produced as per the specifications included in different standards which are issued for structural steels. Structural steels for use at ambient or moderately elevated temperatures are of the following types.

- Carbon and carbon-manganese steels: In these steels the maximum content for alloying elements does not exceed the following: manganese – 1.65 %, silicon – 0.40 % and copper – 0.6 %. The specified minimum of copper does not exceed 0.4 % and also there is no minimum content is specified for other elements to obtain a desired alloying effect.

- High strength low alloy (HSLA) steels: These steels have specified minimum yield strengths greater than 280 Newtons /Sq cm and achieve that strength in hot rolled condition rather than by heat treatment.

- Heat treated high tensile steels: Both Carbon and HSLA steels can be heat treated to provide yield strengths in the range of 350 to 520 Newtons/Sq cm.

- Constructional alloy steels: These steels contain alloying elements in excess of the limits for carbon steels and are heat treated to obtain a combination of high strength and toughness. These are the strongest steels in general structural use with yield strength of 700 Newtons / Sq cm.

- High performance steels: These steels are with enhanced notch toughness and generally used in construction of bridges. These steels are higher in strength, lighter in weight and have greater atmospheric corrosion resistance than conventional steels.

Elements Used in Structural Steels

The following are the important chemical elements which are used in the structural steels.

- Carbon: Carbon is the most important chemical element in the structural steel. Increase in the carbon percentage improves strength but reduces ductility. Hence structural steels have carbon content in the range of 0.15 % to 0.30 %.

- Manganese: Manganese content in structural steels in up to 1.65 % maximum. It has effects similar to those of carbon. Manganese and carbon in structural steels are to be in such combination so as to have the desired properties in the steel.

- Silicon: It is one of the principal deoxidizer for the structural steel. It is the element that is most commonly used for the production of semi killed or fully killed structural steels. Silicon content in structural steels is less than 0.40 %.

- Aluminum: Aluminum is used as deoxidizer in the production of semi killed or fully killed structural steels. It forms a more fine grained crystalline micro structure in the structural steel.

- Phosphorus and sulphur: Both phosphorus and sulphur are usually undesirable elements in the structural steels. Sulphur promotes internal segregation in the steel matrix. Both these elements act to reduce the ductility of the structural steel. There detrimental effect on the steel weldability is also significant. Hence the contents of these elements in the structural steel are limited to less than 0.04 % to 0.05 %.

- Copper: It is a corrosion resistance element and is a primary anti corrosion component in the weathering structural steels. In such steels the copper content should not be less than 0.2 % but ideally it should be in the range of 0.25 % to 0.55 %.

- Niobium: Niobium is the usual strength-enhancing element in HSLA steels. It also has some corrosion resistance properties.

- Chromium: Chromium is present in some structural steels in small amounts. It is usually used to enhance the corrosion resistance of the structural steel and hence often it is used in combination with nickel and copper.

- Molybdenum: It is used in some grades of structural steels to increase the strength of the steel at the higher temperature. It also improves corrosion resistance. It is often used in structural steels in combination with either manganese or vanadium.

- Nickel: Nickel improves the low temperature behaviour of the structural steel by improving the fracture toughness. It also has favorable effect on the corrosion resistance of the structural steel.

- Vanadium: The effect of vanadium is similar to those of manganese, niobium and molybdenum. It helps structural steels to develop a finer crystalline microstructure and have increased fracture toughness.

- Other chemical elements: Some structure steels grades have small amounts of some other alloying elements such as boron, nitrogen and titanium. These minor alloying elements along with the major alloying elements enhance certain capabilities of the structural steels.

Weldability of Sructural Steels

The weldability of structural steels depends on its carbon euivalent (CE) which is defined below.

$$CE = C + Mn/6 + (Cr+Mo+V)/5 + (Ni+Cu)/15$$

Structural steels with carbon equivalent below 0.45 are readily weldable with appropriate procedures. CE of strutural steels greater than 0.45 indicates caution is to be observed during welding of such steels.

Properties of Structural Steels

The following are the important properties of structural steels:

- It has high strength which means that the weight of the structure made of steel is less.

- It has uniform properties which do not change as oppose to concrete.

- Elasticity of structural steel follows Hooke's law accurately.

- It has good ductility due to which structural steels can withstand extensive deformation without failure under high tensile stresses.

- It has good toughness which means that structural steel has got both strength and ductility.

- It has got flexibility which allows extension of existing structures made of structural steels can be done easily.

- Structural steels are non combustible materials but they lose strength when heated sufficiently. When heated to temperatures normally associated with fires, the strength and stiffness of the structural steel are significantly reduced. Structural steels have a critical temperature. This critical temperature is the temperature at which the structural steel cannot safely support its load. Critical temperature is usually the temperature at which its yield stress has been reduced to 60 % of the room temperature yield stress.

- Structural steels can corrode when in contact with water.

- Structural steels can be coated or painted to provide it corrosion and fire resistance.

I-beam

The I-beam, also called the H-beam, wide beam, W-beam, universal beam (UB), and rolled steel joist, is the shape of choice for structural steel builds. The design and structure of the I-beam makes it uniquely capable of handling a variety of loads.

Engineers use I beams widely in construction, forming columns and beams of many different lengths, sizes, and specifications. Understanding the I beam is a basic necessity for the modern civil engineer or construction worker.

The Shape and Structure of the I-beam

The I-beam consists of two horizontal planes, known as flanges, connected by one vertical component, or the web. The shape of the flanges and the web create an "I" or an "H" cross-section. Most I beams use structural steel, but some are made from aluminum. Infra-metal constructions, such as carbon structural steel and high-strength low-alloy structural steel, have different applications – such as building framing, bridges, and general structural purposes.

I beams come in a variety of weights, section depths, flange widths, web thicknesses, and other specifications for different purposes. When ordering I beams, buyers classify them by their material and dimensions. For example, an 11x20 I beam would have an 11-inch depth and a weight of 20 pounds per foot. Builders choose specific sizes of I beams according to the needs of the particular building. A builder has to take many factors into account, such as:

1. Deflection. The builder will choose a thickness to minimize deformation of the beam.

2. Vibration. A certain mass and stiffness are selected to prevent vibrations in the building.

3. Bend. The strength of the I beam's cross-section should accommodate yield stress.

4. Buckling. The flanges are chosen to prevent buckling locally, sideways, or torsionally.

5. Tension. The builder chooses an I beam with a web thickness that won't fail, buckle, or ripple under tension.

The design of the I beam makes it capable of bending under high stress instead of buckling. To achieve this, most of the material in the I beam is located in the regions along the axial fibers – the location that experiences the most stress. Ideal beams have minimal cross-section area, requiring the least amount of material possible while still achieving the desired shape.

Uses of I Beams

I beams have a variety of important uses in the structural steel construction industry. They are often used as critical support trusses, or the main framework, in buildings. Steel I beams ensure a structure's integrity with relentless strength and support. The immense power of I beams reduces the need to include numerous support structures, saving time and money, as well as making the structure more stable. The versatility and dependability of I beams make them a coveted resource to every builder.

I beams are the choice shape for structural steel builds because of their high functionality. The shape of I beams makes them excellent for unidirectional bending parallel to the web. The horizontal flanges resist the bending movement, while the web resists the shear stress. They can take various types of loads and shear stresses without buckling. They are also cost effective, since the "I" shape is an economic design that doesn't use excess steel. With a wide variety of I beam types, there is a shape and weight for virtually any requirement. The versatile functionality of the I-beam is what gives it the alternate name universal beam, or UB.

Steel I Beam Fabrication

When you need I beams for any type of building application, look to steel beam fabrication for fast, efficient, and affordable order fulfillment. Steel beam fabrication takes a lot of experience, knowledge, hard work, and specialized tools to be successful.

Design for Bending

A beam under bending sees high stresses along the axial fibers that are farthest from the neutral axis. To prevent failure, most of the material in the beam must be located in these regions. Comparatively little material is needed in the area close to the neutral axis. This observation is the basis of the I-beam cross-section; the neutral axis runs along the center of the web which can be relatively thin and most of the material can be concentrated in the flanges.

The ideal beam is the one with the least cross-sectional area (and hence requiring the least material) needed to achieve a given section modulus. Since the section modulus depends on the value of the moment of inertia, an efficient beam must have most of its material located as far from the neutral axis as possible. The farther a given amount of material is from the neutral axis, the larger is the section modulus and hence a larger bending moment can be resisted.

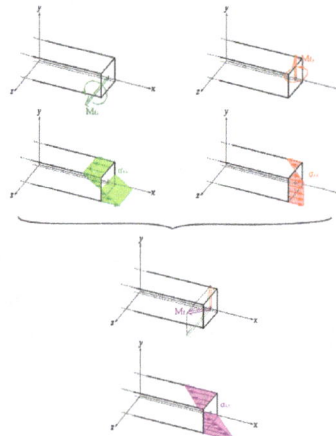

When designing a symmetric I-beam to resist stresses due to bending the usual starting point is the required section modulus. If the allowable stress is σ_{max} and the maximum expected bending moment is M_{max}, then the required section modulus is given by;

$$S = \frac{M_{max}}{\sigma_{max}} = \frac{I}{c}$$

where I is the moment of inertia of the beam cross-section and c is the distance of the top of the beam from the neutral axis.

For a beam of cross-sectional area a and height h, the ideal cross-section would have half the area at a distance $h/2$ above the cross-section and the other half at a distance $h/2$ below the cross-section. For this cross-section

$$I = \frac{ah^2}{4} \; ; \; S = 0.5ah$$

However, these ideal conditions can never be achieved because material is needed in the web for physical reasons, including to resist buckling. For wide-flange beams, the section modulus is approximately

$$S \approx 0.35ah$$

which is superior to that achieved by rectangular beams and circular beams.

Issues

Though I-beams are excellent for unidirectional bending in a plane parallel to the web, they do not perform as well in bidirectional bending. These beams also show little resistance to twisting and undergo sectional warping under torsional loading. For torsion dominated problems, box beams and other types of stiff sections are used in preference to the I-beam.

Wide-flange Steel Materials and Rolling Processes (U.S.)

Rusty riveted steel I-beam

In the United States, the most commonly mentioned I-beam is the wide-flange (W) shape. These beams have flanges in which the planes are nearly parallel. Other I-beams include American Standard (designated S) shapes, in which flange surfaces are not parallel, and H-piles (designated HP), which are typically used as pile foundations. Wide-flange shapes are available in grade ASTM A992, which has generally replaced the older ASTM grades A572 and A36. Ranges of yield strength:

- A36: 36,000 psi (250 MPa)

- A572: 42,000–60,000 psi (290–410 MPa), with 50,000 psi (340 MPa) the most common

- A588: Similar to A572

- A992: 50,000–65,000 psi (340–450 MPa)

Like most steel products, I-beams often contain some recycled content.

The American Institute of Steel Construction (AISC) publishes the Steel Construction Manual for designing structures of various shapes. It documents the common approaches, Allowable Stress Design (ASD) and Load and Resistance Factor Design (LRFD), to create such designs.

Standards

The following standards define the shape and tolerances of I-beam steel sections:

Euronorms

- EN 10024, Hot rolled taper flange I sections – Tolerances on shape and dimensions.

- EN 10034, Structural steel I and H sections – Tolerances on shape and dimensions.

- EN 10162, Cold rolled steel sections – Technical delivery conditions – Dimensional and cross-sectional tolerances

Other

- DIN 1025-5

- ASTM A6, American Standard Beams

- BS 4-1

- IS 808 – Dimensions hot rolled steel beam, column, channel and angle sections

- AS/NZS 3679.1 – Australia and New Zealand standard

Designation and Terminology

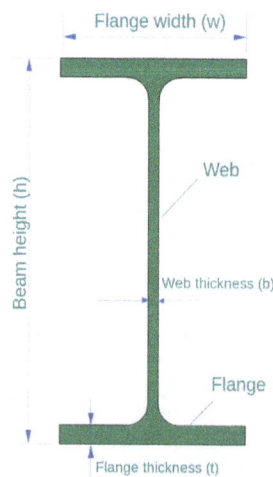

Wide-flange I-beam.

- In the United States, steel I-beams are commonly specified using the depth and weight of the beam. For example, a "W10x22" beam is approximately 10 in (25 cm) in depth (nominal height of the I-beam from the outer face of one flange to the outer face of the other flange) and weighs 22 lb/ft (33 kg/m). Wide flange section beams often vary from their nominal depth. In the case of the W14 series, they may be as deep as 22.84 in (58.0 cm).

- In Canada, steel I-beams are now commonly specified using the depth and weight of the beam in metric terms. For example, a "W250x33" beam is approximately 250 millimetres (9.8 in) in depth (height of the I-beam from the outer face of one flange to the outer face of the other flange) and weighs approximately 33 kg/m (67 lb/yd). I-beams are still available in U.S. sizes from many Canadian manufacturers.

- In Mexico, steel I-beams are called IR and commonly specified using the depth and weight of the beam in metric terms. For example, a "IR250x33" beam is approximately 250 mm (9.8 in) in depth (height of the I-beam from the outer face of one flange to the outer face of the other flange) and weighs approximately 33 kg/m (22 lb/ft).

- In India I-beams are designated as ISMB, ISJB, ISLB, ISWB. ISMB: Indian Standard Medium Weight Beam, ISJB: Indian Standard Junior Beams, ISLB: Indian Standard Light Weight Beams, and ISWB: Indian Standard Wide Flange Beams. Beams are designated as per respective abbreviated reference followed by the depth of section, such as for example "ISMB 450", where 450 is the depth of section in millimetres (mm). The dimensions of these beams are classified as per IS:808 (as per BIS).

- In the United Kingdom, these steel sections are commonly specified with a code consisting of the major dimension (usually the depth)-x-the minor dimension-x-the mass per metre ending with the section type, all measurements being metric. Therefore, a 152x152x23UC would be a column section (UC = universal column) of approximately 152 mm (6.0 in) depth 152 mm width and weighing 23 kg/m (46 lb/yd) of length.

- In Australia, these steel sections are commonly referred to as Universal Beams (UB) or Columns (UC). The designation for each is given as the approximate height of the beam, the type (beam or column) and then the unit metre rate (e.g., a 460UB67.1 is an approximately 460 mm (18.1 in) deep universal beam that weighs 67.1 kg/m (135 lb/yd)).

American Standard Beams AISC

Type	Beam height (in)	Flange width (in)	Web thickness (in)	Flange thickness (in)	Weight (lb/ft)	Cross-section area (in²)	Moment of inertia in torsion (J (cm⁴)
W4x13	4.16	4.06	0.28	0.345	13	3.83	0.151
W5x16	5.01	5	0.24	0.36	16	4.71	0.192
W5x19	5.15	5.03	0.27	0.43	19	5.56	0.316
W6x8.5	5.83	3.94	0.17	0.195	8.5	2.52	0.0333
W6x9	5.9	3.94	0.17	0.215	9	2.68	0.0405
W6x12	6.03	4	0.23	0.28	12	3.55	0.0903
W6x15	5.99	5.99	0.23	0.26	15	4.43	0.101
W6x16	6.28	4.03	0.26	0.405	16	4.74	0.223

Indian Standard Beams ISMB

Type	Beam height (mm)	Flange width (mm)	Web thickness (mm)	Flange thickness (mm)	Weight (kg/m)	Cross-section area (cm²)	Moment of inertia in torsion (J) (cm⁴)
ISMB 80	80	46	3.8	5.2	6.0	7.64	0.712
ISMB 100	100	75	4.0	7.2	11.5	14.6	1.10
ISMB 120	120	70	4.4	6.3	10.4	13.2	1.71
ISMB 140	140	73	4.7	6.9	12.9	16.4	2.54
ISMB 750 × 137	753	263	11.5	17	137	175	137.1
ISMB 750 × 147	753	265	13.2	17	147	188	161.5
ISMB 750 × 173	762	267	14.4	21.6	173	221	273.6
ISMB 750 × 196	770	268	15.6	25.4	196	251	408.9

European Wide Flange Beams HEA and HEB

Type	Beam height (mm)	Flange width (mm)	Web thickness (mm)	Flange thickness (mm)	Weight (kg/m)	Cross-section area (cm²)	Moment of inertia in torsion (J) (cm⁴)
HE 100 A	96	100	5	8	16.7	21.2	5.24
HE 120 A	114	120	5	8	19.9	25.3	5.99
HE 140 A	133	140	5.5	8.5	24,7	31.4	8.13
HE 160 A	152	160	6	9	30.4	38.8	12.19
HE 1000 × 415	1020	304	26	46	415	528.7	2714
HE 1000 × 438	1026	305	26.9	49	437	557.2	3200
HE 1000 × 494	1036	309	31	54	494	629.1	4433
HE 1000 × 584	1056	314	35.6	64	584		

Cellular Beams

Cellular beams are the modern version of the traditional "castellated beam" which results in a beam approximately 40–60% deeper than its parent section. The exact finished depth, cell diameter and cell spacing are flexible. A cellular beam is up to 1.5 times stronger than its parent section and is therefore utilized to create efficient large span constructions.

Steel Channel

Like other hollow sections, steel channel is rolled from steel sheet into C or U shapes. It consists of a wide "web" and two "flanges". The flanges could be parallel or tapered.

Steel channels are extensively used in a lot of applications. There are mainly three types of surface treatments to code with the conditions. Black or non-treatment is not frequently used for the steel will rust easily without any protective layers. Hot-dip galvanization and primer is the common treatments. Zinc coating resists environmental and weather corrosion, while primer does better. You can choose any kind according to your own application.

Metal channel sections are primarily used in industrial, commercial
and architectural construction markets. The channel ceiling grid
can support lights, manufacturing tools or others.

Steel channel has been classified into "C" and "U" types depending on its outlook. Hence there are
two standards to show the channel dimensions - UPE & UPN. UPE is for the C channel steel with
parallel flanges while UPN for the U channel steel with tapered flanges. You can find the size details
in the specific type of channel steel like A36 steel channel. Bespoke channel dimensions are available.

Uses of Channel Steel

Steel channel are one of the most popular parts in construction and manufacturing. Apart from
this, C channel & u channel are also used in our everyday life if you have so much attention to them
like stair stringer. However, owing to its bending axis is not centered on the width of flanges, structural channel steel is not so strong as I beam or wide flange beam.

- Tracks & sliders for machines, doorways, etc.

- Posts and supports for building corners, walls & railings.

- Protective edges for walls.

- Decorative elements for constructions like ceiling channel system.

- Frames or framing material for construction, machines.

Lally Column

A lally column is a steel column designed for structural use. It is very thin (Less then ¼ inch,) leading to the majority of the inside of the column being open space. Normally, a lally column is filled
in with concrete. The concrete in turn helps improve the structural stability of the lally column by
providing compressive resistance. In addition, the concrete helps prevent local buckling in the lally
column.

Together, the lally column with concrete can provide much needed structural support. As a result,
it is mostly used vertically, and can be found in the basement of many homes, providing the necessary vertical support for the first floor wood beams. As a final note, the Lally Column was named
after John Lally, who took out the patent for the design, and used it frequently in his construction
business.

Strengths of the Lally Column

As previously stated, the lally column can provide a great deal of structural stability. In addition to this, the lally column is made out of materials that are easily accessible. Using steel and concrete, the lally column has the added benefit of be cost effective, due to the low cost of the two primary ingredients. On top of this, the lally column is mostly hollow, making it easy to cut with smaller cutting tools. This added versatility makes it a favorite among individuals working on a construction site.

Weaknesses of the Lally Column

A lally column is only as good as the materials used. As a result, poor quality concrete and steel will lead to quick ware and tare. In addition, the lally column design is particularly susceptible to corrosion in moist environments.

Despite these drawbacks, lally columns are still widely used today, providing support for any number of structures. Whether a lally column is right for you or not will depend entirely on your situation.

Engineered Wood

Engineered wood is manufactured from scraps of lumber and byproducts, such as saw dust, that have been reformed using heat, glue and pressure to make a usable solid-wood alternative. In an effort to meet the demands for a construction material that is lower costing but equally durable, the forest industry engineered several types of wood-based board products. Engineered woods are intended to be used as a substrate, which means they are covered with a laminate or wood veneer. This is a construction material that has grown in popularity because of advancements in the engineered wood's quality being produced today.

Particle board is a commonly known engineered wood used in the manufacturing of many kitchen cabinets and countertops, but it is not the only one. Another equally popular and arguably more durable engineered wood is plywood. Fiberboard and flakeboard are the other popular versions on the market today. All of the previously mentioned board types use the similar methods to manufacture the board, but the type of raw material wood used will vary the final texture and strength between them. Processes that are achieving higher densities and the advancement into the inclusion of fire retardant materials are helping to improve the reputation of engineered woods.

A building using EWPs sequesters carbon and can enhance the performance of the building, especially in energy use. Wood has natural thermal properties, which greatly reduce heat transfer. While many materials transmit heat easily, wood does not due to its structure. Imagine looking at a box of straws from the end — that is what wood looks like when magnified about 200 times. These hollow cellulose straws both reduce the wood's density and provide insulation properties.

Types of Products

Engineered wood products in a Home Depot store.

Plywood

Plywood, a wood structural panel, is sometimes called the original engineered wood product. Plywood is manufactured from sheets of cross-laminated veneer and bonded under heat and pressure with durable, moisture-resistant adhesives. By alternating the grain direction of the veneers from layer to layer, or "cross-orienting", panel strength and stiffness in both directions are maximized. Other structural wood panels include oriented strand board and structural composite panels.

Densified Wood

Densified wood is made by using a mechanical hot press to compress wood fibers and increase the density by a factor of three. This increase in density is expected to enhance the strength and stiffness of the wood by a proportional amount. Early studies confirmed this result with a reported increase in mechanical strength by a factor of three.

Chemically Densified Wood

More recent studies have combined chemical process with traditional mechanical hot press methods to increase density and thus mechanical properties of the wood. In these methods, chemical processes break down lignin and hemicellulose that is found naturally in wood. Following dissolution, the cellulose strands that remain are mechanically hot compressed. Compared to the three-fold increase in strength observed from hot pressing alone, chemically processed wood has been shown to yield an 11-fold improvement. This extra strength comes from hydrogen bonds formed between the aligned cellulose nanofibers.

The densified wood possessed mechanical strength properties on par with steel used in building construction, opening the door for applications of densified wood in situations where regular strength wood would fail. Environmentally, wood requires significantly less carbon dioxide to produce than steel and acts as a source for carbon sequestration.

Fibreboard

Medium-density fibreboard, is made by breaking down hardwood or softwood residuals into wood fibres, combining it with wax and a resin binder, and forming panels by applying high temperature and pressure.

Particle Board

Particle board is manufactured from wood chips, sawmil shavings, or even sawdust, and a synthetic resin or other suitable binder, which is pressed and extruded. Oriented strand board, also known as flakeboard, waferboard, or chipboard, is similar but uses machined wood flakes offering more strength. Particle board is cheaper, denser and more uniform than conventional wood and plywood and is substituted for them when cost is more important than strength and appearance. A major disadvantage of particleboard is that it is very prone to expansion and discoloration due to moisture, particularly when it is not covered with paint or another sealer.

Oriented Strand Board

Oriented strand board (OSB) is a wood structural panel manufactured from rectangular-shaped strands of wood that are oriented lengthwise and then arranged in layers, laid up into mats, and bonded together with moisture-resistant, heat-cured adhesives. The individual layers can be cross-oriented to provide strength and stiffness to the panel. However, most OSB boards are delivered with more strength in one direction. The wood strands in the outmost layer on each side of the board are normally aligned into the strongest direction of the board. Arrows on the product will often identify the strongest direction of the board (When bought in most cases the height (The longest dimension) of the board). Produced in huge, continuous mats, OSB is a solid panel product of consistent quality with no laps, gaps or voids.

OSB is delivered in various dimensions, strengths and levels of water resistance.

Laminated Timber

Glued laminated timber (glulam) is composed of several layers of dimensional timber glued together with moisture-resistant adhesives, creating a large, strong, structural member that can be used as vertical columns or horizontal beams. Glulam can also be produced in curved shapes, offering extensive design flexibility.

Laminated Veneer

Laminated veneer lumber (LVL) is produced by bonding thin wood veneers together in a large billet. The grain of all veneers in the LVL billet is parallel to the long direction. The resulting product features enhanced mechanical properties and dimensional stability that offer a broader range in product width, depth and length than conventional lumber. LVL is a member of the structural composite lumber (SCL) family of engineered wood products that are commonly used in the same structural applications as conventional sawn lumber and timber, including rafters, headers, beams, joists, rim boards, studs and columns.

Cross Laminated

Cross-Laminated Timber (CLT) is a versatile multi-layered panel made of lumber. Each layer of boards is placed cross-wise to adjacent layers for increased rigidity and strength. CLT can be used for long spans and all assemblies, e.g. floors, walls or roofs. CLT has the advantage of faster construction times as the panels are manufactured and finished off site and supplied ready to fit and screw together as a flat pack assembly project.

Parallel Strand

Parallel strand lumber (PSL) consists of long veneer strands laid in parallel formation and bonded together with an adhesive to form the finished structural section. A strong, consistent material, it has a high load carrying ability and is resistant to seasoning stresses so it is well suited for use as beams and columns for post and beam construction, and for beams, headers, and lintels for light framing construction. PSL is a member of the structural composite lumber (SCL) family of engineered wood products.

Laminated Strand

Laminated strand lumber (LSL) and oriented strand lumber (OSL) are manufactured from flaked wood strands that have a high length-to-thickness ratio. Combined with an adhesive, the strands are oriented and formed into a large mat or billet and pressed. LSL and OSL offer good fastener-holding strength and mechanical connector performance and are commonly used in a variety of applications, such as beams, headers, studs, rim boards, and millwork components. These products are members of the structural composite lumber (SCL) family of engineered wood products. LSL is manufactured from relatively short strands—typically about 1 foot long—compared to the 2 foot to 8 foot long strands used in PSL.

Finger Joint

The finger joint is made up of short pieces of wood combined to form longer lengths and is used in doorjambs, mouldings and studs. It is also produced in long lengths and wide dimensions for floors.

Beams

I-joists and wood I-beams are "I"-shaped structural members designed for use in floor and roof construction. An I-joist consists of top and bottom flanges of various widths united with webs of various depths. The flanges resist common bending stresses, and the web provides shear performance. I-joists are designed to carry heavy loads over long distances while using less lumber than a dimensional solid wood joist of a size necessary to do the same task. As of 2005, approximately half of all wood light framed floors were framed using I-joists.

Trusses

Roof trusses and floor trusses are structural frames relying on a triangular arrangement of webs and chords to transfer loads to reaction points. For a given load, long wood trusses built from smaller pieces of lumber require less raw material and make it easier for AC contractors, plumbers, and electricians to do their work, compared to the long 2x10s and 2x12s traditionally used as rafters and floor joists.

Transparent Wood Composites

Transparent wood composites are new composites made at the laboratory scale that combine transparency and stiffness. They are not available yet on the market.

Advantages

Engineered wood products are used in a variety of ways, often in applications similar to solid wood products. Engineered wood products may be preferred over solid wood in some applications due to certain comparative advantages.

- Because engineered wood is man-made, it can be designed to meet application-specific performance requirements. Required shapes and dimension do not drive source tree requirements (length or width of the tree)

- Engineered wood products are versatile and available in a wide variety of thicknesses, sizes, grades, and exposure durability classifications, making the products ideal for use in unlimited construction, industrial and home project application.

- Engineered wood products are designed and manufactured to maximize the natural strength and stiffness characteristics of wood. The products are very stable and some offer greater structural strength than typical wood building materials.

- Glued laminated timber (glulam) has greater strength and stiffness than comparable dimensional lumber and, pound for pound, is stronger than steel.

- Some engineered wood products offer more design options without sacrificing structural requirements.

- Engineered wood panels are easy to work with using ordinary tools and basic skills. They can be cut, drilled, routed, jointed, glued, and fastened. Plywood can be bent to form curved surfaces without loss of strength. And large panel size speeds construction by reducing the number of pieces to be handled and installed.

- Engineered wood products make more efficient use of wood. They can be made from small pieces of wood, wood that has defects or underutilized species.

- Wooden trusses are competitive in many roof and floor applications, and their high strength-to-weight ratios permit long spans offering flexibility in floor layouts.

- Engineered wood is felt to offer structural advantages for home construction.

- Sustainable design advocates recommend using engineered wood, which can be produced from relatively small trees, rather than large pieces of solid dimensional lumber, which requires cutting a large tree.

Disadvantages

- Some products may burn more quickly than solid lumber.

- They require more primary energy for their manufacture than solid lumber.

- The adhesives used in some products may be toxic. A concern with some resins is the release of formaldehyde in the finished product, often seen with urea-formaldehyde bonded products.

- Cutting and otherwise working with some products can expose workers to toxic compounds.

- Some engineered wood products, such as those specified for interior use, may be weaker and more prone to humidity-induced warping than equivalent solid woods. Most particle and fiber-based boards are not appropriate for outdoor use because they readily soak up water.

Properties

Plywood and OSB typically have a density of 550 - 650 kg/m³ (35 to 40 pounds per cubic foot). For example, 1 cm (3/8") plywood sheathing or OSB sheathing typically has a weight of 1 - 1.2 kg/m² (1.0 to 1.2 pounds per square foot.) Many other engineered woods have densities much higher than OSB.

Lamella

The lamella is the face layer of the wood that is visible when installed. Typically, it is a sawn piece of timber. The timber can be cut in three different styles: flat-sawn, quarter-sawn, and rift-sawn. Keep in mind that each cut will give the board a different final appearance.

Core/Substrate

1. Wood ply construction ("sandwich core"): Uses multiple thin plies of wood adhered together. The wood grain of each ply runs perpendicular to the ply below it. Stability is attained from using thin layers of wood that have little to no reaction to climatic change. The wood is further stabilized due to equal pressure being exerted lengthwise and widthwise from the plies running perpendicular to each other.

2. Finger core construction: Finger core engineered wood floors are made of small pieces of milled timber that run perpendicular to the top layer (lamella) of wood. They can be 2-ply or 3-ply, depending on their intended use. If it is three ply, the third ply is often plywood that runs parallel to the lamella. Stability is gained through the grains running perpendicular to each other, and the expansion and contraction of wood is reduced and relegated to the middle ply, stopping the floor from gapping or cupping.

3. Fibreboard: The core is made up of medium or high density fibreboard. Floors with a fibreboard core are hygroscpoic and must never be exposed to large amounts of water or very high humidity - the expansion caused from absorbing water combined with the density of the fibreboard, will cause it to lose its form. Fibreboard is less expensive than timber and can emit higher levels of harmful gases due to its relatively high adhesive content.

4. An engineered flooring construction which is popular in parts of Europe is the hardwood lamella, softwood core laid perpendicular to the lamella, and a final backing layer of the same noble wood used for the lamella. Other noble hardwoods are sometimes used for the back layer but must be compatible. This is thought by many to be the most stable of engineered floors.

Aesthetics

Engineered wood flooring is mainly industrially fabricated in the form of straight edged boards, with milled jointing profiles to provide for interconnecting of the boards. Such manufacturing is most cost efficient but leaves an industrial looking surface. In nature, no straight lines exist; therefore there is a rising trend to modify the visual appearance to imitate it. In recent years, numerous producers have been taking on the challenge of adding more natural aesthetics.

Adhesives

The types of adhesives used in engineered wood include:

Urea-formaldehyde resins (UF): most common, cheapest, and not waterproof.

Phenol formaldehyde resins (PF): yellow/brown, and commonly used for exterior exposure products.

Melamine-formaldehyde resins (MF): white, heat and water resistant, and often used in exposed surfaces in more costly designs.

Polymeric Methylene diphenyl diisocyanate (pMDI) or polyurethane (PU) resins: expensive, generally waterproof, and does not contain formaldehyde, notoriously more difficult to release from platens and engineered wood presses.

A more inclusive term is *structural composites*. For example, fiber cement siding is made of cement and wood fiber, while cement board is a low-density cement panel, often with added resin, faced with fiberglass mesh.

Health Concerns

While formaldehyde is an essential ingredient of cellular metabolism in mammals, studies have linked prolonged inhalation of formaldehyde gases to cancer. Engineered wood composites have been found to emit potentially harmful amounts of formaldehyde gas in two ways: unreacted free formaldehyde and chemical decomposition of resin adhesives. When exorbitant amounts of formaldehyde are added to a process, the excess will not have any additive to bond with and may seep from the wood product over time. Cheap urea-formaldehyde (UF) adhesives are largely responsible for degraded resin emissions. Moisture degrades the weak UF molecules, resulting in potentially harmful formaldehyde emissions. McLube offers release agents and platen sealers designed for those manufacturers who use reduced-formaldehyde UF and melamine-formaldehyde adhesives. Many oriented strand board (SB) and plywood manufacturers use phenol-formaldehyde (PF) because phenol is a much more effective additive. Phenol forms a water-resistant bond with formaldehyde that will not degrade in moist environments. PF resins have not been found to pose significant health risks due to formaldehyde emissions. While PF is an excellent adhesive, the engineered wood industry has started to shift toward polyurethane binders like pMDI to achieve even greater water-resistance, strength, and process efficiency. pMDIs are also used extensively in the production of rigid polyurethane foams and insulators for refrigeration. pMDIs outperform other resin adhesives, but they are notoriously difficult to release and cause buildup on tooling surfaces.

Other Fixations

Some engineered products such as CLT Cross Laminated Timber can be assembled without the use of adhesives using mechanical fixing. These can range from profiled interlocking jointed boards, proprietary metal fixings, nails or timber dowels (Brettstapel - single layer or CLT).

Purlin

Purlins are supported either by rafters or the walls of the building. They are most commonly used in metal buildings, though they sometimes replace closely spaced rafters in wood frame structures.

Purlin

The purlins of a roof support the weight of the roof deck. The roof deck is the wood panel, ply board, or metal sheeting that creates the surface of the roof. When made of wood, it is usually covered with some sort of weatherproofing and sometimes an insulation material.

Several kinds of purlins exist. They are divided into categories based on the material from which they are made and their shape. Different purlins are used for different purposes, including structural support of walls or floors. Purlin is important because without it, there's no frame for the sheeting on the roof to rest on, making purlins critical to the structure of the roof.

1. Purlin Material

a. Wood Purlin

Wood Purlin is good for use with the fibre cement sheeting. The wood purlin and sheeting combine well to ensure that the room below is breathable and can safely store whatever you need to be kept safe in the room, from livestock to grain or other organic materials.

However, being made from wood, purlins can rot. Besides, the main problem with wood is it being dry when it goes up. Therefore, it is best dried before installation. Moreover, the moisture can add significantly to the weight leading to sag.

Wood purlin

b. Steel Purlin

Steel Purlin is a direct replacement for wood purlin. They are light weight, dimensionally stable, accurate and straight. They expand and contract reasonably in extreme temperature changes.

Characteristics	Steel purlin	Wood purlin
Cost	Generally cheaper	More expensive
Recyclability	Recyclable, thus "greener"	Less recyclable
Assembling	Quicker to be assembled	Slower to be assembled
Weight	Lighter than wood structures of the same size	Heavier than steel structures of the same size
Insect proof	Not affected by termites or other insects that feed on wood	Affected by termites
Life time	Last much longer than wood in most applications	Shorter life time than steel
Fire resistance	Greater fire resistance	Flammable

Steel purlin is usually made of cold-formed steel that is thin enough to put screws through. Cold-formed steel is made by rolling or pressing thin sheets of steel into the desired shape. It is less expensive for the manufacturer than hot-rolled steel and is also easier to work with. Though cold-formed steel is stronger than hot-formed steel, it is more likely to break when under pressure rather than bend.

Purlins are manufactured from hot dipped galvanized steel with a coating, in line with other common lightweight steel structural building products. This gives good protection in most exposed internal environments. Run off from, or contact with, materials which are incompatible with zinc should be avoided.

To protect the purlin, they also apply a layer of paint outside them. Zinc and paint in combination (synergistic effect) produce a corrosion protection approximately 2 times the sum of the corrosion protection that each alone would provide.

Steel purlin

2. Purlin Types

There are three types of steel purlins:

C Purlin (Cee Purlin)

Cee purlin or C purlin is shaped like a squared-off letter C. C Purlins are horizontal structures that are used to support the load from the roof deck or the sheathing. The plane surface of this purling on one side has made it a preferred material for cladding due to its easy installation on concrete structures or steel. Cee purlin should be light in weight and perfect for simple span construction.

C Purlins are often used for structural support in walls and as floor joists in addition to roofs. Cee purlins may also be used to form braces, ties, or columns in sheds.

Features

- Optimum quality
- Easy installation
- High tensile strength
- Abrasion & corrosion resistance

Advantages

- Ability to span length

- Purlin erection is easier than others
- Fast to erect and easy handling
- No side drilling/cutting required
- Assured dimensions and straightness
- High durability, versatility and uniform quality
- Low transportation cost due to decreased weight
- Close tolerances on sectional sizes due to process of cold roll forming

C purlin (left) and Z purlin (right)

Z Purlin (Zed Purlin)

Zed purlin or Z purlin is shaped like a letter Z. Z purlin shape allows the purlin to overlap with others at the joints. This gives Zed purlins the potential to be much stronger than C purlins. They are mainly used in walls or for large roofing projects.

Z Purlins are made using cold-formed or rolled sheets for supporting roof. The flexible shape of these beams facilitates various designs solutions. These purlins are extensively used in huge roofing solutions such as godowns, workshops, industrials sheds and many more. The range is known for saving up to 50% on structural sheet in comparison with hot rolled angles. Z purlin should be crisp and clean in design and do not allow the scope of inaccurate lengths.

Advantages

- Ability to span length
- Saving in Steel up to 40%
- Fast to erect and easy handling
- No side drilling/cutting required
- Assured dimensions and straightness
- Purlin erection is easier than others
- Saving in construction cost up to 30%
- High durability, versatility and uniform quality

- Low transportation cost due to reduced weight

- Close tolerances on sectional dimensions owing to process of cold roll forming

- Saving up to 35-40% in weight and 20% in cost when compared to hot rolled purlins

Rectangular Hollow Section (RHS)

Rectangular hollow section (RHS) is a type of purlin often used in roofs where the support structure will be visible when construction is complete. For example, decks and covered patios often use this type of purlin. As the bar is hollow, caps are welded to the ends of the purlin to keep moisture from getting inside and corroding the metal. The rectangular shape gives the roof the same aesthetic quality of a roof supported by wooden beams.

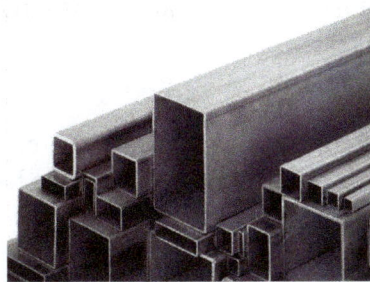
Rectangular hollow section (RHS)

3. Purlin vs Rafter

Rafter are also the basic member of any roof structure. Together with purlin, they are the load transmitting members of the steel roof. They transmit the live load, dead load, wind load (which is considerable in large steel structures with low permeability) and other loads acting on them to the roof truss situated below them which eventually transfers the load to the columns and then eventually to the foundation.

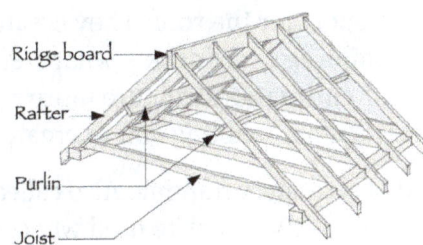
Purlin and Rafter

So basically, purlin and rafter are like two ways reinforcements of the roof. Purlins are one which are parallel to the ridge line or we can say they run along the span of the roof while rafters are perpendicular (plan) to the ridge line of the roof truss.

The roof truss is supported on the columns and the rafter is supported on the roof truss upon which the purlins are bolted/welded and upon the purlins lies the final covering material which may be asbestos sheet or cement asbestos sheet. Also note here that the rafters are present only where the truss are not present to support the purlins.

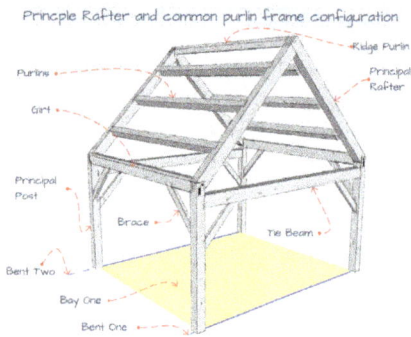

Principle rafter and common purlin frame configuration

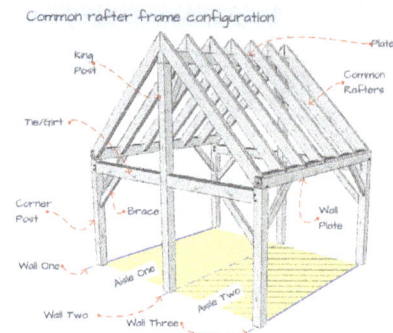

Common rafter frame configuration

For easier distinction, the rafters are orientated vertically along the slope of the roof. Purlins are orientated horizontally, providing support to the rafters, usually at mid-height.

4. Purlin and Girt

Secondary framing is an important component of many pre-engineered metal buildings. Also referred to as "secondary structural", this type of framing runs in between primary framing elements, creating a structure-within-a-structure, much like cross-beams in a wooden building.

The purpose of secondary framing is to distribute loads from the building's surfaces to the main framing and the foundation. Secondary framing can add longitudinal support that helps resist wind and earthquakes. And it can provide lateral bracing for compression flanges that are part of the primary framing, increasing overall frame capacity.

Secondary framing components are known as girts and purlins, and they work like this:

- Girts provide additional support for walls: They work in conjunction with columns and wall panels to support vertical load, improving both strength and stability. They also help attach and support wall cladding.

- Purlins provide additional support for the roof: They create a horizontal "diaphragm" that supports the weight of your building's roof deck – whatever material you use for the roof itself. They also help make your entire roof structure more rigid. Because they add mid-span support, purlins allow longer spans, enabling you to create a wider building.

- Eave struts are another kind of secondary framing: Also called eave girts or eave purlins, these are essentially a combination of the two. They're used where sidewalls intersect with the roof, using a top flange that helps support the roof and a "web" that helps supports the walls.

Secondary framing comes in two configurations, CEE and ZED. They're shaped on a bending press, to create a web with two flanges. They come in a variety of sizes; for instance, purlins can run over 30 feet in length.

Girts, purlins and eave struts are almost always made of cold-formed steel. It's more affordable and easier to work with, but it also presents some structural stability issues that must be considered as part of your metal building framing options and overall design. In particular, local or distortional buckling or lateral displacement can occur, in which portions of the compression flange, web or connectors can buckle or shift from their initial position.

Problems can occur under extreme stress or even under relatively low stress if conditions are just right. However, you shouldn't consider these engineering issues to be detractors if you're considering a metal building. Additional stability or support can come not only from girts, purlins and eave struts but also from additional stiffeners.

Exactly how many and what size secondary framing elements your building may require will depend on your building's dimensions, primary framing system and how you plan to use the building as well as other factors. Metal building company can explain the nuances in detail and guide you toward the right decision.

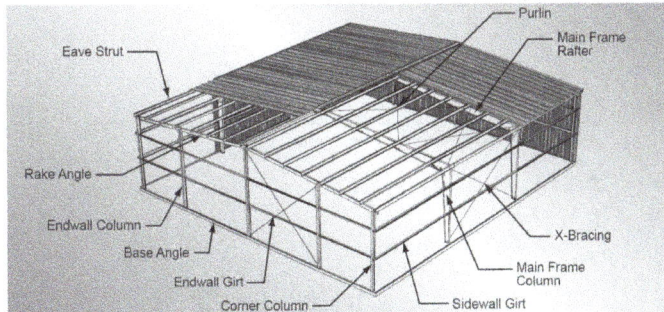

Purlin and Girt

5. Purlin Roof

Roof purlin needs no introduction to anyone in the construction industry. In their design life, purlins are subjected to dead load (e.g. self weight of sheeting materials and accessories), live load (e.g. during maintenance services and repairs), and environmental loads (e.g. wind and snow load). Therefore, a purlin should be adequately strong to withstand the loads it will encounter during its design life and should not sag in an obvious manner thereby giving the roof sheeting an undulating and/or unpleasant appearance. This post will be focusing on design of steel purlin using cold formed sections.

6. Purlin Span

The span referred to is the distance between centers of the cleat bolts at each end of the purlin. Each span type represents a complete purlin run system and recognizes that using separate component parts (e.g. internal span, end span) is not a valid procedure.

a. Single Purlin Span

Single purlin span is the span that is simply supported by means of bolting the web of the purlin to a cleat or other rigid structure. Under these conditions bridging does not influence inward capacities, but outward capacities vary dependent on the number of rows of bridging.

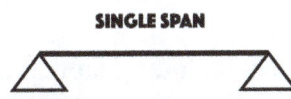

SINGLE SPAN

b. Double Purlin Span

Double purlin span are simply supported at each end and in the center. They may comprise only one purlin over the full length or two purlins lapped together over the central support to provide continuity. Both inward and outward capacities are influenced by bridging in double spans.

DOUBLE UNLAPPED

DOUBLE LAPPED

c. Continuous Purlin Span

Continuous purlin span are simply supported at each end and at a series of equally spaced intermediate supports. These tables are for spans in which purlins are lapped over each support, where the lap length is 15% of the span. Tables are given for 5 or more spans. For outwards loads on equal continuous spans the bridging shown is required in the end spans. One less row of bridging can be used for internal spans if inwards capacity, the recommended minimum bridging and practical spacing requirements allow. For inwards loads the required bridging in the table applies to all spans.

CONTINUOUS LAPPED (5 SPAN SHOWN)

7. Purlin Accessories

a. Purlin Laps

It is generally recommended a lap length of 15% of the span. Where span lengths are unequal (e.g. reduced end spans) each purlin should have 7.5% of each adjacent span added rather than 7.5% of that purlin's span.

Lap lengths of less than 10% (or 5% on any side of a support) may not provide full structural continuity and may also suffer from local failures not considered by this method. They must therefore be considered beyond the scope of this manual.

b. Purlin Cleats

Single cleats are used in most situations, including lapped Z purlins. Double cleats are generally only used where successive purlins (usually unlapped) are butted together. Double cleats could also be used in applications with a high reaction load, to reduce bolt stresses. In this situation, additional care would be needed in hole detailing.

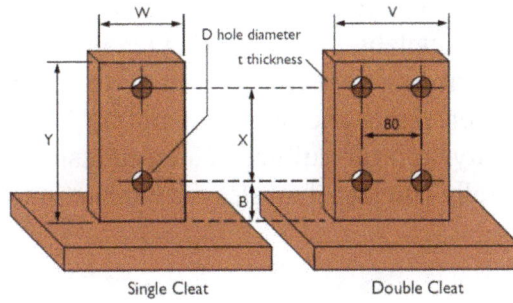

Single Cleat   Double Cleat

8. Purlin Design

a. Purlin Spacing

By default, purlin sections assume the slope of the roof they are supporting. The spacing of purlins usually call for careful arrangement, in the sense that it should follow the nodal pattern of the supporting trusses. What I mean in this regard is that purlins should be placed at the nodes of trusses and not on the members themselves so as not to induce secondary bending and shear forces in the members of the truss. Furthermore, if manual analysis is employed to analyze a truss loaded in such manner, such secondary stresses cannot be captured since we normally assume pinned connections. Cold formed Z (Zed) and C (Cee) sections are normally specified for purlins in steel structures.

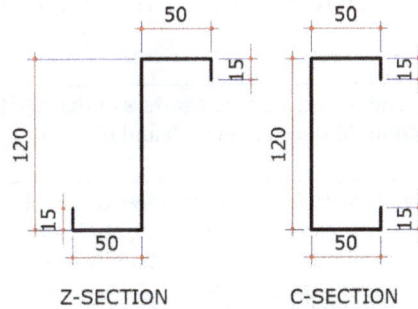

Z-SECTION   C-SECTION

Z-Section and C-Section.

As compared with thicker hot rolled shapes, they normally offer the advantages of lightness, high strength and stiffness, easy fabrication and installations, easy packaging and transportation etc. The connection of purlins can be sleeved or butted depending on the construction method adopted.

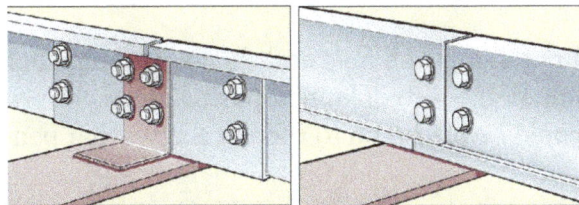

Sleeved connection and Butted connection.

In terms of arrangement, we can have single spans with staggered sleeved/butt arrangement, single/double span with staggered sleeve arrangement, double span butt joint system, and single span butt joint system. The choice of the arrangement to be adopted can depend on the supply length of the sections as readily available in the market, the need to avoid wasteful offcuts, the loading and span of the roof, the arrangement of the rafters etc. Therefore, the roof designer must plan from start to finish. However, single and double span butt joint system are the most popular in Nigeria, due to their simplicity, and the culture of adopting shorter roof spans in the country. However, they are less structurally efficient than sleeved connections.

b. Purlin Bridging

Bridging provides resistance to purlin rotation during the installation of roof and wall sheeting. For this reason, a maximum bridging (or bridging to cleat) spacing of 20 x purlin depth, but no greater than 4000, is recommended. Failure to do so can lead to misaligned fastenings, causing additional stresses on the fasteners and roof sheeting. Excessive purlin rotation can be a safety hazard during construction. It is therefore recommended that at least one row of bridging be used in each purlin span. Span/bridging configurations which exceed these recommendations are shown on the left of or above the red line. However, in instances where sheeting is successfully installed outside of this recommendation the published values are valid for structural purposes.

9. Purlin Installation

Purlins are installed horizontally under metal roofs. They are installed on top of the roof rafters with a felt underlayment or vapor barrier installed on top. Purlins are 2 by 4 feet and are installed much like metal roofing. They give added support to the roof and also provide a nailing surface for the end panels and drip edge.

Step 1	Run a tape measure from end to end along the rafters of the roof to determine how many purlins will be needed for installation. Measure the width and height of the roof and record the measurements.
Step 2	Snap a chalk line horizontally across the roof two feet down from the top. Lay the first purlin at the ridge of the roof down to the chalk line beginning at either corner. Fasten the purlin with 16d common nails into each vertical rafter. Insert two equally spaced nails into the rafter. The rafters are generally spaced every 16 inches on center.
Step 3	Set the second purlin horizontally right next to the first and install it the same way. Continue down the row until the first row is complete. Cut the last purlin to size with tin snips, if necessary.
Step 4	Move two feet down the roof and snap a horizontal chalk line. Install the second row of purlins just like the first, cutting the end piece to fit. Snap the third chalk line two more feet down and continue down the roof until it is covered with purlins. Inspect the area to verify that no nails are sticking up and that all the purlins are secure to the rafters.

a. Purlin Laps

Purlin laps must be bolted in the top web hole and the lower flange holes at both ends of the lap as shown below. Bolting only in the web of lapped purlins does not provide full structural continuity and excessive loads could be placed on to roofing screws that penetrate both purlins within a lapped region.

b. Purlin Flying Bracing

If the lower web hole in a lap is used for attaching fly bracing ensure that an additional bolt is used.

c. Purlin Bridging

Design Station Bridging can be installed either up or down the roof slope but cannot be mixed within a bridging run. However, as the starting and finishing components will be different, the direction of fixing must be established at the design/procurement phase. Girt bridging must not exceed its compressive capacity. Where more than one row is to be installed always complete the bridging for each girt before commencing on the next (i.e. do not complete one row of bridging before starting the next).

d. Welding

Purlin Installation

The welding or hot cutting of purlins, girts or bridging is not recommended. The heat produced in welding will affect the material properties of the high yield strength cold-formed steel used in purlins. In many instances considerable stress concentrations are likely to arise, even with good quality welding. In addition, welding will locally remove the protective coating, leading to a potential reduction in durability.

10. Purlin Handling and Storage

Roof Purlins must be kept dry during storage as water present between close stacked sections will cause premature corrosion. If they become wet, they should be separated and stacked openly to allow for ventilation to dry the surface.

The installation of Purlins can be hazardous and will require an adequate safety plan be in place prior to handling or installing of these products. All rigging, scaffolding and safety equipment must comply with the relevant codes, Australian standards and statutory requirements. It is recommended that good trade practice be followed such as that outlined in Australian Standards AS3828-1998 (Guidelines for the erection of building steelwork) and HB39 (Installation code for metal roof and wall cladding). Normally, purlins are not designed to be walked on unless fully covered by correctly installed roofing materials or the correct grade of safety mesh. The manufacturing or delivery process may result in oil or grease adhering to these purlins which could increase the potential hazard. Handling of this product must be carried out using a correctly supervised crane or appropriate lifting device. Safety harnesses must always be used during installation of purlins when working off the ground, and under no circumstances must any body weight be placed on bridging, or on purlins or girts that have not been fully bolted into position and with the correct bridging installed. Bolts must be the correct size and grade, all progressively fully tightened during installation. Laps must be bolted in the outer web hole (closest to the sheeting), and the inner flange holes provided.

11. Purlin Supplier

Last but not least, you have to choose the best supplier for you to order purlin. A reputable supplier with strong customer base can ensure the purlin's quality to a certain extent. Other than reputation, there are some elements that make the best purlin supplier.

a. Production Capacity

Production capacity of the manufacturer decides how long it takes you to receive the purlin. The higher the production capacity is, the shorter time you have to wait. It depends on their factory size and technology they applied. In general, big manufacturer have higher production capacity.

In case of steel purlin, it's essential to find a manufacturer which has production capacity of around 500MTS per day.

b. Production Process

The production is recommended to be closed from raw material to final product. If so, the manufacturer can completely manage the quality of the purlin. If they are not steelmakers and need other steel supplier, they cannot be sure about the exact material and production process. What will happen if your purlin are rusty after 3 months being used and when you contact your supplier, they blame it for the steelmaker. Moreover, it results in higher price because the products' prices include more transportation costs and benefits.

c. Technology

Each production stage should use separate and modern production techniques. Modern machinery and advanced technology applied will save energy and reduce production costs. There are global standards for the quality management system and product quality such as ISO, Standards Australia, ASTM, JIS G, ... Make sure you ask the supplier about their quality certification before you order anything.

d. Consultancy

Professional consultancy is very important when you merely know about purlin and steel. They will help you decide the purlin sizes and span according to your place's weather condition. They had better provide free consultancy and it is convenient with 24/7 service.

Professional consultancy is very important when you merely know about purlin and steel.

e. Professional Staffs

If you are looking for a foreign supplier, they are needed to be familiar with exporting activities.

You had better work with knowledgeable and dynamic staffs, which will push the process going much faster.

f. After-sales service

After buying purlin, you still need assistance in setting and maintenance. You will be in trouble if your supplier disappears right after sales because there will be no one to guarantee the quality of the products and take responsibility when they are faulty.

Lumber

Lumber is used for structural purposes primarily. Lumber also produces furniture, pulp, paper and other composites like plywood and particleboard. It is also used as fuel while cooking and heating.

Lumber is classified as either hardwood or softwood. Softwood comprises pine fir, spruce, cedar, hemlock, cypress, redwood and other conifer trees. Conifer trees have exposed seeds that are surrounded cones. Most conifer trees are evergreen. Some of the famous softwood trees in the U.S. are southern yellow pine and Douglas fir.

Softwood is easy to saw and nail, which is why it is ideally suited to make buildings, furniture and paper.

Hardwood consists of trees like oak and maple wood. These trees are mostly deciduous and broad-leaved.

Hardwood has many colors and patterns and is used traditionally for producing commercial products, pallets, high-end furniture, cabinets, flooring, tables, etc.

Rough lumber, consisting of both hardwood and softwood, is used to make furniture and other items that require some reshaping. Finished lumber comes in standard sizes and is used primarily in the construction industry for flooring.

Lumber is abundant in nature, though conservationists have pointed out that lumber is a finite source.

Lumber has been used for thousands of years to construct houses, carts, and several items of daily use. It is a widely traded commodity, and is used in a number of industrial applications. Its primary consumer is the construction industry that uses it as a raw material to build homes.

Lumber may suffer from some defects that may affect the price of the supply. These defects include chip marks, diagonal grains, torn grains and wane during processing or splinters due to improper seasoning. Lumber is also prone to fungi attacks (in high-moisture environments), termites, carpenter ants and wood boring beetles.

Each tree produces different type of lumber, which varies in grain, knots and natural wear. Lumber is usually categorized according to thickness or lengths. Lumber thickness is measured in inches, while lengths can be from 8 feet to 20 feet. The lumber futures contract is based on lumber of dimensions 2 inch by 4 inches. This is the most widely produced variety in America.

In order to produce lumber, trees are first felled and pruned of all leaves and branches. Then, they are cut into logs, transporter by trucks to sawmills for processing.

At the mills, logs are sawed into different sizes depending on existing demand and prices. The lumber is dried in kilns so that all moisture is removed from the wood. Next it is smoothed and graded.

Ways of Using Timber

Softwood Lumber is used for residential home construction and remodelling of homes. Nearly half of the United States' softwood is used up by the construction industry for boards and planks. As per the National Association of Home Builders, a home of around 2,400 square feet would need around 14,400 board feet of softwood. Our pallets, cabinets and flooring come from hardwood lumber.

Housing starts, published monthly by the U.S. Department of Commerce are popularly used to indicate residential construction activity, which further indicates the demand for lumber.

Currently, China imports around half of its softwood demand from North America.

As per global forecasts, Asia will become the largest consumer of lumber in the next five years, especially due to the growth of wood pellets and chips as biomass source. Brazil, China and Russia will also increase their output.

Remanufactured Lumber

Remanufactured lumber is the result of secondary or tertiary processing/cutting of previously milled lumber. Specifically, it is lumber cut for industrial or wood-packaging use. Lumber is cut by ripsaw or resaw to create dimensions that are not usually processed by a primary sawmill.

Resawing is the splitting of 1-inch through 12-inch hardwood or softwood lumber into two or more thinner pieces of full-length boards. For example, splitting a ten-foot 2×4 into two ten-foot 1×4s is considered resawing.

Plastic Lumber

Structural lumber may also be produced from recycled plastic and new plastic stock. Its introduction has been strongly opposed by the forestry industry. Blending fiberglass in plastic lumber enhances its strength, durability, and fire resistance. Plastic fiberglass structural lumber can have a "class 1 flame spread rating of 25 or less, when tested in accordance with ASTM standard E 84," which means it burns slower than almost all treated wood lumber.

Conversion of Wood Logs

Logs are converted into timber by being sawn, hewn, or split. Sawing with a rip saw is the most common method, because sawing allows logs of lower quality, with irregular grain and large knots, to be used and is more economical. There are various types of sawing:

- Plain sawn (flat sawn, through and through, bastard sawn) – A log sawn through without adjusting the position of the log and the grain runs across the width of the boards.

- Quarter sawn and rift sawn – These terms have been confused in history but generally mean lumber sawn so the annual rings are reasonably perpendicular to the sides (not edges) of the lumber.

- Boxed heart – The pith remains within the piece with some allowance for exposure.

- Heart center – the center core of a log.

- Free of heart center (FOHC) – A side-cut timber without any pith.

- Free of knots (FOK) – No knots are present.

Dimensional Lumber

Dimensional lumber is lumber that is cut to standardized width and depth, specified in inches. Carpenters extensively use dimensional lumber in framing wooden buildings. Common sizes include 2×4 (also two-by-four and other variants, such as four-by-two in Australia, New Zealand, and the UK), 2×6, and 4×4. The length of a board is usually specified separately from the width and depth. It is thus possible to find 2×4s that are four, eight, and twelve feet in length. In Canada and the United States, the standard lengths of lumber are 6, 8, 10, 12, 14, 16, 18, 20, 22 and 24 feet (1.83, 2.44, 3.05, 3.66, 4.27, 4.88, 5.49, 6.10, 6.71 and 7.32 meters). For wall framing, "stud" or "precut" sizes are available, and are commonly used. For an eight-, nine-, or ten-foot ceiling height, studs are available in 92 $\frac{5}{8}$ inches (235 cm), 104 $\frac{5}{8}$ inches (266 cm), and 116 $\frac{5}{8}$ inches (296 cm). The term "stud" is used inconsistently to specify length; where the exact length matters, one must specify the length explicitly.

A common 2×4 board

Historical Chinese Construction

Under the prescription of the Method of Construction issued by the Southern Song government in the early 12th century, timbers were standardized to eight cross-sectional dimensions. Regardless of the actual dimensions of the timber, the ratio between width and height was maintained at 1:1.5. Units are in Song Dynasty inches (3.12 cm).

Class	height	width	uses
1st	9	6	great halls 11 or 9 bays wide
2nd	8.25	5.5	great halls 7 or 5 bays wide
3rd	7.5	5	great halls 5 or 3 bays wide or halls 7 or 5 bays wide
4th	7.2	4.8	great halls 3 bays wide or halls 5 bays wide
5th	6.6	4.4	great halls 3 small bays wide or halls 3 large bays wide
6th	6	4	pagodas and small halls
7th	5.25	3.2	pagodas and small great halls
8th	4.5	3	small pagodas and ceilings

Timber smaller than the 8th class were called "unclassed". The width of a timber is referred to as one "timber", and the dimensions of other structural components were quoted in multiples of "timber"; thus, as the width of the actual timber varied, the dimensions of other components were easily calculated, without resorting to specific figures for each scale. The dimensions of timbers in similar application show a gradual diminution from the Sui Dyansty (580~618) to the modern era; a 1st class timber during the Sui was reconstructed as 15×10 (Sui Dynasty inches, or 2.94 cm).

North American Softwoods

The length of a unit of dimensional lumber is limited by the height and girth of the tree it is milled from. In general the maximum length is 24 ft (7.32 m). Engineered wood products, manufactured by binding the strands, particles, fibers, or veneers of wood, together with adhesives, to form composite materials, offer more flexibility and greater structural strength than typical wood building materials.

Pre-cut studs save a framer much time, because they are pre-cut by the manufacturer for use in 8-, 9-, and 10-ft (2.44, 2.74 and 3.05 m) ceiling applications, which means the manufacturer has removed a few inches or centimetres of the piece to allow for the sill plate and the double top plate with no additional sizing necessary.

In the Americas, *two-bys* (2×4s, 2×6s, 2×8s, 2×10s, and 2×12s), named for traditional board thickness in inches, along with the 4×4 (89 mm × 89 mm), are common lumber sizes used in modern construction. They are the basic building blocks for such common structures as balloon-frame or platform-frame housing. Dimensional lumber made from softwood is typically used for construction, while hardwood boards are more commonly used for making cabinets or furniture.

Lumber's nominal dimensions are larger than the actual standard dimensions of finished lumber. Historically, the nominal dimensions were the size of the green (not dried), rough (unfinished) boards that eventually became smaller finished lumber through drying and planing (to smooth the wood). Today, the standards specify the final finished dimensions and the mill cuts the logs to whatever size it needs to achieve those final dimensions. Typically, that rough cut is smaller than the nominal dimensions because modern technology makes it possible and it uses the logs more efficiently. For example, a "2×4" board historically started out as a green, rough board actually 2 by 4 inches (51 mm × 102 mm). After drying and planing, it would be smaller, by a nonstandard amount. Today, a "2×4" board starts out as something smaller than 2 inches by 4 inches and not specified by standards, and after drying and planing is reliably 1 ½ by 3 ½ inches (38 mm × 89 mm).

North American softwood dimensional lumber sizes

| Nominal | Actual | | Nominal | Actual | | Nominal | Actual | | Nominal | Actual | | Nominal | Actual | |
inches	inches	mm	inches	inches	mm	inches	inches	mm	inches	inches	mm	inches	inches	mm
1 × 2	$3/4 \times 1\frac{1}{2}$	19 × 38	2 × 2	$1\frac{1}{2} \times 1\frac{1}{2}$	38 × 38									
1 × 3	$3/4 \times 2\frac{1}{2}$	19 × 64	2 × 3	$1\frac{1}{2} \times 2\frac{1}{2}$	38 × 64									
1 × 4	$3/4 \times 3\frac{1}{2}$	19 × 89	2 × 4	$1\frac{1}{2} \times 3\frac{1}{2}$	38 × 89	4 × 4	$3\frac{1}{2} \times 3\frac{1}{2}$	89 × 89						
1 × 5	$3/4 \times 4\frac{1}{2}$	19 × 114												
1 × 6	$3/4 \times 5\frac{1}{2}$	19 × 140	2 × 6	$1\frac{1}{2} \times 5\frac{1}{2}$	38 × 140	4 × 6	$3\frac{1}{2} \times 5\frac{1}{2}$	89 × 140	6 × 6	$5\frac{1}{2} \times 5\frac{1}{2}$	140 × 140			
1 × 8	$3/4 \times 7\frac{1}{4}$	19 × 184	2 × 8	$1\frac{1}{2} \times 7\frac{1}{4}$	38 × 184	4 × 8	$3\frac{1}{2} \times 7\frac{1}{4}$	89 × 184				8 × 8	$7\frac{1}{4} \times 7\frac{1}{4}$	184 × 184
1 × 10	$3/4 \times 9\frac{1}{4}$	19 × 235	2 × 10	$1\frac{1}{2} \times 9\frac{1}{4}$	38 × 235									
1 × 12	$3/4 \times 11\frac{1}{4}$	19 × 286	2 × 12	$1\frac{1}{2} \times 11\frac{1}{4}$	38 × 286									

Early standards called for green rough lumber to be of full nominal dimension when dry. However, the dimensions have diminished over time. In 1910, a typical finished 1-inch (25 mm) board was $^{13}/_{16}$ in (21 mm). In 1928, that was reduced by 4%, and yet again by 4% in 1956. In 1961, at a meeting in Scottsdale, Arizona, the Committee on Grade Simplification and Standardization agreed to what is now the current U.S. standard: in part, the dressed size of a 1-inch (nominal) board was fixed at $^{3}/_{4}$ inch; while the dressed size of 2 inch (nominal) lumber was *reduced* from $1\,^{5}/_{8}$ inch to the current $1\,^{1}/_{2}$ inch.

Dimensional lumber is available in green, unfinished state, and for that kind of lumber, the nominal dimensions are the actual dimensions.

Grades and Standards

Individual pieces of lumber exhibit a wide range in quality and appearance with respect to knots, slope of grain, shakes and other natural characteristics. Therefore, they vary considerably in strength, utility, and value.

The longest board in the world is in Poland and measures 36.83 metres (about 120 ft 10 in) long.

The move to set national standards for lumber in the United States began with publication of the American Lumber Standard in 1924, which set specifications for lumber dimensions, grade, and moisture content; it also developed inspection and accreditation programs. These standards have changed over the years to meet the changing needs of manufacturers and distributors, with the goal of keeping lumber competitive with other construction products. Current standards are set by the American Lumber Standard Committee, appointed by the U.S. Secretary of Commerce.

Design values for most species and grades of visually graded structural products are determined in accordance with ASTM standards, which consider the effect of strength reducing characteristics, load duration, safety and other influencing factors. The applicable standards are based on results of tests conducted in cooperation with the USDA Forest Products Laboratory. Design Values for Wood Construction, which is a supplement to the ANSI/AF&PA National Design Specification for Wood Construction, provides these lumber design values, which are recognized by the model building codes.

Canada has grading rules that maintain a standard among mills manufacturing similar woods to assure customers of uniform quality. Grades standardize the quality of lumber at different levels and are based on moisture content, size, and manufacture at the time of grading, shipping, and unloading by the buyer. The National Lumber Grades Authority (NLGA) is responsible for writing, interpreting and maintaining Canadian lumber grading rules and standards. The Canadian Lumber Standards Accreditation Board (CLSAB) monitors the quality of Canada's lumber grading and identification system.

Attempts to maintain lumber quality over time have been challenged by historical changes in the timber resources of the United States – from the slow-growing virgin forests common over a century ago to the fast-growing plantations now common in today's commercial forests. Resulting declines in lumber quality have been of concern to both the lumber industry and consumers and have caused increased use of alternative construction products.

Machine stress-rated and machine-evaluated lumber is readily available for end-uses where high strength is critical, such as trusses, rafters, laminating stock, I-beams and web joints. Machine grading measures a characteristic such as stiffness or density that correlates with the structural properties of interest, such as bending strength. The result is a more precise understanding of the strength of each piece of lumber than is possible with visually graded lumber, which allows designers to use full-design strength and avoid overbuilding.

In Europe, strength grading of rectangular sawn timber (both softwood and hardwood) is done according to EN-14081 and commonly sorted into classes defined by EN-338. For softwoods the common classes are (in increasing strength) C16, C18, C24 and C30. There are also classes specifically for hardwoods and those in most common use (in increasing strength) are D24, D30, D40, D50, D60 and D70. For these classes, the number refers to the required 5th percentile bending strength in Newtons per square millimetre. There are other strength classes, including T-classes based on tension intended for use in glulam.

- C14, used for scaffolding and formwork.

- C16 and C24, general construction.

- C30, prefab roof trusses and where design requires somewhat stronger joists than C24 can offer. TR26 is also a common trussed rafter strength class in long standing use in the UK.

- C40, usually seen in glulam.

Grading rules for African and South American sawn timber have been developed by ATIBT according to the rules of the Sciages Avivés Tropicaux Africains (SATA) and is based on clear cuttings – established by the percentage of the clear surface.

North American Hardwoods

In North America, market practices for dimensional lumber made from hardwoods varies significantly from the regularized standardized 'dimension lumber' *sizes* used for sales and specification of softwoods – hardwood boards are often sold totally rough cut, or machine planed only on the two (broader) face sides. When Hardwood Boards are also supplied with planed faces, it is usually both by random widths of a specified thickness (normally matching milling of softwood dimensional lumbers) and somewhat random lengths. But besides those older (traditional and normal) situations, in recent years some product lines have been widened to also market boards in standard stock sizes; these usually retail in big-box stores and using only a relatively small set of specified lengths; in all cases hardwoods are sold to the consumer by the board-foot (144 cubic inches or 2,360 cubic centimetres), whereas that measure is not used for softwoods at the retailer (to the cognizance of the buyer).

North American hardwood dimensional lumber sizes Nominal (rough-sawn size)	S1S (surfaced on one side)	S2S (surfaced on two sides)
$\frac{1}{2}$ in	$\frac{3}{8}$ in (9.5 mm)	$\frac{5}{16}$ in (7.9 mm)
$\frac{5}{8}$ in	$\frac{1}{2}$ in (13 mm)	$\frac{7}{16}$ in (11 mm)
$\frac{3}{4}$ in	$\frac{5}{8}$ in (16 mm)	$\frac{9}{16}$ in (14 mm)
1 in or $\frac{4}{4}$ in	$\frac{7}{8}$ in (22 mm)	$\frac{13}{16}$ in (21 mm)
1 $\frac{1}{4}$ in or $\frac{5}{4}$ in	1 $\frac{1}{8}$ in (29 mm)	1 $\frac{1}{16}$ in (27 mm)
1 $\frac{1}{2}$ in or $\frac{6}{4}$ in	1 $\frac{3}{8}$ in (35 mm)	1 $\frac{5}{16}$ in (33 mm)
2 in or $\frac{8}{4}$ in	1 $\frac{13}{16}$ in (46 mm)	1 $\frac{3}{4}$ inches (44 mm)
3 in or $\frac{12}{4}$ in	2 $\frac{13}{16}$ in (71 mm)	2 $\frac{3}{4}$ in (70 mm)
4 in or $\frac{16}{4}$ in	3 $\frac{13}{16}$ in (97 mm)	3 $\frac{3}{4}$ in (95 mm)

Also in North America, hardwood lumber is commonly sold in a "quarter" system, when referring to thickness; 4/4 (four quarter) refers to a 1-inch-thick (25 mm) board, 8/4 (eight quarter) is a 2-inch-thick (51 mm) board, etc. This "quarter" system is rarely used for softwood lumber; although softwood decking is sometimes sold as 5/4, even though it is actually one-inch thick (from milling 1/8th inch off each side in a motorized planing step of production). The "quarter" system of reference is a traditional (cultural) North American lumber industry nomenclature used specifically to indicate the thickness of rough sawn hardwood lumber.

In recent years architects, designers, and builders have begun to use the "quarter" system in specifications as a vogue of insider knowledge, though the materials being specified are finished lumber, thus conflating the separate systems and causing confusion.

Hardwoods cut for furniture are cut in the fall and winter, after the sap has stopped running in the trees. If hardwoods are cut in the spring or summer the sap ruins the natural color of the timber and decreases the value of the timber for furniture.

Engineered Lumber

Engineered lumber is lumber created by a manufacturer and designed for a certain structural purpose. The main categories of engineered lumber are:

- Laminated veneer lumber (LVL) – LVL comes in $1\frac{3}{4}$ inch thicknesses with depths such as $9\frac{1}{2}$, $11\frac{7}{8}$, 14, 16, 18, and 24 inches, and are often doubled or tripled up. They function as beams to provide support over large spans, such as removed support walls and garage door openings, places where dimensional lumber is insufficient, and also in areas where a heavy load is bearing from a floor, wall or roof above on a somewhat short span where dimensional lumber is impractical. This type of lumber is compromised if it is altered by holes or notches anywhere within the span or at the ends, but nails can be driven into it wherever necessary to anchor the beam or to add hangers for I-joists or dimensional lumber joists that terminate at an LVL beam.

- Wooden I-joists – sometimes called "TJI", "Trus Joists" or "BCI", all of which are brands of wooden I-joists, they are used for floor joists on upper floors and also in first floor conventional foundation construction on piers as opposed to slab floor construction. They are engineered for long spans and are doubled up in places where a wall will be aligned over them, and sometimes tripled where heavy roof-loaded support walls are placed above them. They consist of a top and bottom chord or flange made from dimensional lumber with webbing in-between made from oriented strand board (OSB) (or, latterly, steel mesh forms which allow passage of services without cutting). The webbing can be removed up to certain sizes or shapes according to the manufacturer's or engineer's specifications, but for small holes, wooden I-joists come with "knockouts", which are perforated, pre-cut areas where holes can be made easily, typically without engineering approval. When large holes are needed, they can typically be made in the webbing only and only in the center third of the span; the top and bottom chords lose their integrity if cut. Sizes and shapes of the hole, and typically the placing of a hole itself, must be approved by an engineer prior to the cutting of the hole and in many areas, a sheet showing the calculations made by the engineer must be provided to the building inspection authorities before the hole will be approved. Some I-joists are made with W-style webbing like a truss to eliminate cutting and to allow ductwork to pass through.

Freshly cut logs showing sap running from beneath bark.

- Finger-jointed lumber – solid dimensional lumber lengths typically are limited to lengths of 22 to 24 feet, but can be made longer by the technique of "finger-jointing" by using small solid pieces, usually 18 to 24 inches long, and joining them together using finger joints and glue to produce lengths that can be up to 36 feet long in 2×6 size. Finger-jointing also is predominant in precut wall studs. It is also an affordable alternative for non-structural hardwood that will be painted (staining would leave the finger-joints visible). Care is taken during construction to avoid nailing directly into a glued joint as stud breakage can occur.

- Glulam beams – created from 2×4 or 2×6 stock by gluing the faces together to create beams such as 4×12 or 6×16. As such, a beam acts as one larger piece of lumber – thus eliminating the need to harvest larger, older trees for the same size beam.

- Manufactured trusses – trusses are used in home construction as a pre-fabricated replacement for roof rafters and ceiling joists (stick-framing). It is seen as an easier installation and a better solution for supporting roofs than the use of dimensional lumber's struts and purlins as bracing. In the southern U.S. and elsewhere, stick-framing with dimensional lumber roof support is still predominant. The main drawbacks of trusses are reduced attic space, time required for engineering and ordering, and a cost higher than the dimensional lumber needed if the same project were conventionally framed. The advantages are significantly reduced labor costs (installation is faster than conventional framing), consistency, and overall schedule savings.

Various Pieces and Cuts

- Square and rectangular forms: Plank, slat, batten, board, lath, *strapping* (typically $3/_{4 \text{ in}} \times 1 1/_{2 \text{ in}}$), *cant* (A partially sawn log such as sawn on two sides or squared to a large size and later resawn into lumber. A *flitch* is a type of cant with wane on one or both sides). Various pieces are also known by their uses such as post, beam, (girt), stud, rafter, joist, sill plate, wall plate.

- Rod forms: pole, (dowel), stick (staff, baton)

Timber Piles

In the United States, pilings are mainly cut from southern yellow pines and Douglas firs. Treated pilings are available in Chromated copper arsenate retentions of 0.60, 0.80 and 2.50 pounds per cubic foot (9.6, 12.8 and 40.0 kg/m³) if treatment is required.

Defects in Lumber

Defects occurring in lumber are grouped into the following four divisions:

Conversion

During the process of converting timber to commercial form the following defects may occur:

- Chip mark: this defect is indicated by the marks or signs placed by chips on the finished surface of timber.

- Diagonal grain: improper sawing of timber.

- Torn grain: when a small depression is made on the finished surface due to falling of some tool.

- Wane: presence of original rounded surface in the finished product.

Defects due to Fungi and Animals

Fungi attack timber when these conditions are all present:

- The timber moisture content is above 25% on a dry-weight basis
- The environment is sufficiently warm
- Oxygen (O_2) is present

Wood with less than 25% moisture (dry weight basis) can remain free of decay for centuries. Similarly, wood submerged in water may not be attacked by fungi if the amount of oxygen is inadequate.

Fungi timber defects:

- Blue stain
- Brown rot
- Dry rot
- Heart rot
- Sap stain
- Wet rot
- White rot

Following are the insects and molluscs, which are usually responsible for the decay of timber:

- Woodboring beetles
- Marine borers (Barnea similis)
- Teredos (Teredo navalis)
- Termites
- Carpenter ants
- Carpenter bees

Natural Forces

There are two main natural forces responsible for causing defects in timber: abnormal growth and rupture of tissues. Rupture of tissue includes cracks or splits in the wood called "shakes". "Ring shake", "wind shake", or "ring failure" is when the wood grain separates around the growth rings either while standing or during felling. Shakes may reduce the strength of a timber and the appearance thus reduce lumber grade and may capture moisture, promoting decay. Eastern hemlock is known for having ring shake. A "check" is a crack on the surface of the wood caused by the outside of a timber shrinking as it seasons. Checks may extend to the pith and follow the grain. Like shakes, checks can hold water promoting rot. A "split" goes all the way through a timber. Checks and splits occur more frequently at the ends of lumber because of the more rapid drying in these locations.

Seasoning

The seasoning of lumber is typically either kiln- or air-dried. Defects due to seasoning are the main cause of splits, bowing and honeycombing.

Durability and Service Life

Under proper conditions, wood provides excellent, lasting performance. However, it also faces several potential threats to service life, including fungal activity and insect damage – which can be avoided in numerous ways. Section 2304.11 of the International Building Code addresses protection against decay and termites. This section provides requirements for non-residential construction applications, such as wood used above ground (e.g., for framing, decks, stairs, etc.), as well as other applications.

There are four recommended methods to protect wood-frame structures against durability hazards and thus provide maximum service life for the building. All require proper design and construction:

- Controlling moisture using design techniques to avoid decay.

- Providing effective control of termites and other insects.

- Using durable materials such as pressure treated or naturally durable species of wood where appropriate.

- Providing quality assurance during design and construction and throughout the building's service life using appropriate maintenance practices.

Moisture Control

Wood is a hygroscopic material, which means it naturally absorbs and releases water to balance its internal moisture content with the surrounding environment. The moisture content of wood is measured by the weight of water as a percentage of the oven-dry weight of the wood fiber. The key to controlling decay is controlling moisture. Once decay fungi are established, the minimum moisture content for decay to propagate is 22 to 24 percent, so building experts recommend 19 percent as the maximum safe moisture content for untreated wood in service. Water by itself does not harm the wood, but rather, wood with consistently high moisture content enables fungal organisms to grow.

The primary objective when addressing moisture loads is to keep water from entering the building envelope in the first place, and to balance the moisture content within the building itself. Moisture control by means of accepted design and construction details is a simple and practical method of protecting a wood-frame building against decay. For applications with a high risk of staying wet, designers specify durable materials such as naturally decay-resistant species or wood that has been treated with preservatives. Cladding, shingles, sill plates and exposed timbers or glulam beams are examples of potential applications for treated wood.

Controlling Termites and Other Insects

For buildings in termite zones, basic protection practices addressed in current building codes include (but are not limited to) the following:

- Grading the building site away from the foundation to provide proper drainage.

- Covering exposed ground in any crawl spaces with 6-mil polyethylene film and maintaining at least 12 to 18 inches (300 to 460 mm) of clearance between the ground and the bottom of framing members above (12 inches to beams or girders, 18 inches to joists or plank flooring members).

- Supporting post columns by concrete piers so that there is at least 6 inches (150 mm) of clear space between the wood and exposed earth.

- Installing wood framing and sheathing in exterior walls at least eight inches above exposed earth; locating siding at least six inches from the finished grade.

- Where appropriate, ventilating crawl spaces according to local building codes.

- Removing building material scraps from the job site before backfilling.

- If allowed by local regulation, treating the soil around the foundation with an approved termiticide to provide protection against subterranean termites.

Preservatives

To avoid decay and termite infestation, untreated wood is separated from the ground and other sources of moisture. These separations are required by many building codes and are considered necessary to maintain wood elements in permanent structures at safe moisture content for decay protection. When it is not possible to separate wood from the sources of moisture, designers often rely on preservative-treated wood.

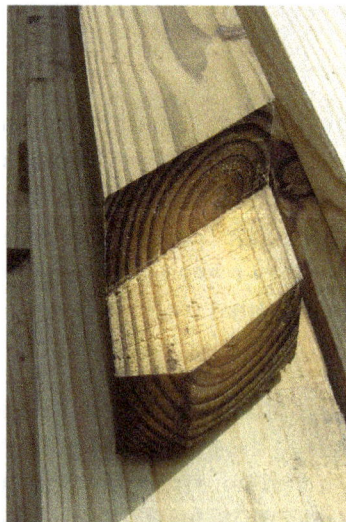

Special fasteners are used with treated lumber because of the corrosive chemicals used in its preservation process.

Wood can be treated with a preservative that improves service life under severe conditions without altering its basic characteristics. It can also be pressure-impregnated with fire-retardant chemicals that improve its performance in a fire. One of the early treatments to "fireproof lumber", which retard fires, was developed in 1936 by the Protexol Corporation, in which lumber is heavily treated with salt. Wood does not deteriorate simply because it gets wet. When wood breaks down, it is because an organism is eating it. Preservatives work by making the food source inedible to these organisms. Properly preservative-treated wood can have 5 to 10 times the service life of untreated wood. Preserved wood is used most often for railroad ties, utility poles, marine piles, decks, fences and other outdoor applications. Various treatment methods and types of chemicals are available, depending on the attributes required in the particular application and the level of protection needed.

There are two basic methods of treating: with and without pressure. Non-pressure methods are the application of preservative by brushing, spraying or dipping the piece to be treated. Deeper, more thorough penetration is achieved by driving the preservative into the wood cells with pressure. Various combinations of pressure and vacuum are used to force adequate levels of chemical into the wood. Pressure-treating preservatives consist of chemicals carried in a solvent. Chromated copper arsenate, once the most commonly used wood preservative in North America began being phased out of most residential applications in 2004. Replacing it are amine copper quat and copper azole.

All wood preservatives used in the United States and Canada are registered and regularly re-examined for safety by the U.S. Environmental Protection Agency and Health Canada's Pest Management and Regulatory Agency, respectively.

Ancient Construction Works

Timber was used as a dominant building material in most of the ancient temples of Kerala and coastal Karnataka of India.

Environmental Effects of Lumber

Green building minimizes the impact or "environmental footprint" of a building. Wood is a major building material that is renewable and replenishable in a continuous cycle. Studies show manufacturing wood uses less energy and results in less air and water pollution than steel and concrete. However, demand for lumber is blamed for deforestation.

Residual Wood

The conversion from coal to biomass power is a growing trend in the United States.

The United Kingdom, Uzbekistan, Kazakhstan, Australia, Fiji, Madagascar, Mongolia, Russia, Denmark, Switzerland and Swaziland governments all support an increased role for energy derived from biomass, which are organic materials available on a renewable basis and include residues and/or byproducts of the logging, sawmilling and papermaking processes. In particular, they view it as a way to lower greenhouse gas emissions by reducing consumption of oil and gas while supporting the growth of forestry, agriculture and rural economies. Studies by the U.S. government have found the country's combined forest and agriculture land resources have the power to sustainably supply more than one-third of its current petroleum consumption.

Biomass is already an important source of energy for the North American forest products industry. It is common for companies to have cogeneration facilities, also known as combined heat and power, which convert some of the biomass that results from wood and paper manufacturing to electrical and thermal energy in the form of steam. The electricity is used to, among other things, dry lumber and supply heat to the dryers used in paper-making.

Timber Framing

Timber framing is a distinctive style of building construction in which heavy timbers frame the structure instead of more slender dimensional lumber (for example, 2 x 6-in.). Timber framing was a building practice used throughout the world until toughly 1900 when the demand for cheap, fast housing brought dimensional lumber to the construction forefront. In the 1970s, craftsman revived the timber framing tradition in the United States and have ushered the design style into the modern era.

One of the most defining elements of a timber frame is its unique joinery. Heavy timber is joined together via mortises and tenons, and then secured with wooden pegs.

What are the differences among dimensionally built, log homes, post and beam, and timber framing?

Dimensionally built structures (sometimes called stick built) are framed with slender dimensional lumber—lumber in preset sizes that are readily available at lumberyards.

Log homes and structures are built of logs stacked horizontally, forming the walls.

In post and beam structures, upright posts support horizontal beams. These may be built of logs (round) or timber (milled to square). Post and beam structures are sometimes made of timber that is held together by metal brackets.

Timber framing is a specialized version of timber post and beam that is built like furniture, using wood joinery such as mortise and tenon, held in place with wooden pegs.

As designs become more intricate and code requirements more stringent, the distinction between some of these common terms becomes blurred. For example, timber frames may require engineered connectors in some joints. These connectors can be hidden inside the joint instead of attached to the timber surface, preserving the traditional timber frame appearance while making use of non-traditional technologies. Also, hybrid structures are prevalent, where timber framing and stick building are each used in the construction of different parts of a building.

Today, there are active timber framing communities in Australia, Canada, Denmark, France, Germany, United Kingdom, United States, and Japan. People from several other countries continue timber framing as individuals. Apprenticeship programs exist in France, Germany, United Kingdom, and the United States.

Raising

In timber framing, all of the structural timbers for a building are prepared ahead of time—sized, planed, and joinery cut, all according to shop drawings. The elements of each cross-section, or bent, are test-fitted; then, when all cutting and testing is done, the bents are assembled and raised from horizontal to vertical one by one. As each bent is raised, it is joined onto the bent already erect by horizontal beams. Then, usually, roof timbers are placed on top of the frame.

This raising can take place by hand—a hand rising—for smaller or historically authentic frames, through the use of pike poles, ropes, and people power. Gin poles and A-frames can help raise a frame. Or cranes can lift bents and timbers into place. Some raisings use a mix of methods.

Joinery

Joinery is what ties timbers together, in traditional timber framing. The ends of timbers are carved out so that they fit together like puzzle pieces. A hole about an inch in diameter is drilled right through the joint, and a wooden peg is pounded in to hold the joint together.

The universe of possible joints is quite large and complex. Common joints include mortise and tenon, dovetail, tying joint, scarf joint, and lap joint. There are many variations and combinations of these and other types of joinery.

SIP

A SIP is a structural insulated panel. SIPs can surround an entire timber frame, leaving all the majestic timber visible in the interior and providing an airtight barrier. A SIP is composed of an insulating foam core (expanded polystyrene or extruded polystyrene) sandwiched between two structural facings, typically oriented strand board. The result is a building system that is extremely strong, energy efficient, and cost-effective.

Built-up Infill Wall

Some builders create their own infill. A typical built-up infill system relies on a groove or channel cut into all faces of the timber in which the plane of the wall lies. On the inside face of the inserted cement board or plywood panel, light framing is installed to receive wood, cement board, or gypsum wallboard for the interior finish of the building. Lathe and plaster may all be applied to the light framing.

Insulation may consist of a hard foam board applied either externally to the plywood panel, or internally between the cement board and light framing.

Generally, the spaces in between the light framing are also insulated, either with a spray insulation, cellulose, or fiber insulation.

Glued Laminated Timber

Glulam is short for glue-laminated timber. It is a versatile and innovative construction material used widely in commercial as well as residential projects. Glulam is made with multiple layers of solid-wood lumber bonded together with high-strength adhesive to form a single structural unit. Builders often refer to glulam beams and other members simply as "glulams."

Glulam Construction

Glue-laminated timber is an engineered wood product, meaning it is made from wood but is machined and assembled to precise specifications to create a predictable, dimensionally stable material. Other common engineered wood products include plywood, oriented strandboard (OSB), and laminated veneer lumber (LVL). A glulam beam looks like a stack of 2x4s (or larger lumber) glued together on their broadsides. This is in contrast to other engineered members, such as LVL and Microlam beams, which have very thin layers of wood glued side by side and look like thick plywood.

Glulam members are sold in several standard widths and lengths and can be custom-fabricated to meet almost any design specifications. They are commonly used for large curved or arching members to build vaulted roofs, domes, and even bridges. Glulam offers superior strength and stiffness over dimensional lumber and it is stronger than steel, pound for pound. Connections for glulams are typically made with bolts or steel dowels and steel plates.

Common Uses of Glulam

Glulam members can be used in a wide variety of applications and for both indoor and outdoor projects. Common forms of glulam members include:

- Straight beams, including lintels, purlins, ridge beams, and floor beams
- Columns, including round, square, and complex sections
- Tied rafters
- Trusses
- Tied arches
- Arched bridge supports
- Curved beams

Glulam Strength Classification

Glulam members are classified by specific strength properties and are given a rating from a stress classification system. The first part of the rating is the reference bending design value, or its flexural loading. For example, a 24F signifies that the member has 2,400 pounds per square inch (psi) of flexural loading. The second part of the rating is the corresponding modulus of elasticity value of the glulam member. For example, a 24F-1.8E indicates a reference bending design value of 2,400 psi and a modulus of elasticity of 1.8×10^6 psi.

Glulam Grades

Glulam material comes in four different appearance grades, as listed in American National Standards Institute (ANSI) A190.1 :

1. Framing: Framing appearance grade is the common choice for homebuilding and other areas where glulam is specified and will be combined with dimension lumber. This grade is recommended only for use in concealed areas.

2. Industrial: Industrial-grade appearance for glulam material is recommended for areas where the aesthetics are not a major concern. Under this appearance grade, the glulam is finished slightly better than with the framing grade, but it is not an aesthetic product. It should be used in areas not visible to the general public. This appearance grade shows some wood imperfection on its surfaces, such as knots and voids.

3. Architectural: When glulam is going to be used as a facade material or exposed element, architectural-grade appearance is highly recommended. This grade offers a high-quality finished product, where wooden voids and imperfections are filled or treated to provide a smoother, more attractive surface.

4. Premium: This grade of glulam is available only through special order and usually is reserved for special situations or predetermined areas, where a high concentration of people is expected. Premium-grade glulam offers the smoothest surfaces for the highest-quality finished product.

Advantages of Glulam

Glulams are not only strong, cost-effective, and highly customizable; they're also resource-efficient because they are made with relatively small pieces of lumber to create a sizable wood member that you could only otherwise get from large, old-growth timber. Glulams offer many benefits to designers and builders:

- Versatile use as roof and floor beams, columns, bracing, decking, and other structural components
- Eco-friendly material with very low formaldehyde levels
- Capable of creating unsupported spans of over 500 feet
- Reduced transportation and handling costs
- Easy installation and surface repair
- Customizable to fit special needs
- Standard sizes available immediately
- Produced in well-managed forests and certified to PEFC standards
- Good fire resistance; can outlast steel beams under the same fire conditions
- Manufactured to precise dimensions

Handling and Storage Tips

Glulam members should be handled carefully to prevent any damage or reduction in their structural capacity. It is recommended that fabric slings be used when lifting to prevent surface scratches. Members should be stored vertically, if possible, and always should be protected from the weather by a covering of plastic sheeting. Unless glulam material is intended for exposed areas, it should be protected from outdoor exposure until it is ready to be installed.

Laminated Veneer Lumber

Laminated Veneer Lumber is a high-strength engineered wood product made from veneers bonded together under heat and pressure. It is used for permanent structural applications including beams and rafters.

Laminated Veneer Lumber (LVL) is a high-strength engineered wood product used primarily for structural applications. It is comparable in strength to solid timber, concrete and steel and is manufactured by bonding together rotary peeled or sliced thin wood veneers under heat and pressure. LVL was developed in the 1970s and is today used for permanent structural applications including beams, lintels, purlins, truss chords and formwork. LVL can be used wherever sawn timber is used however one of the main advantages is that it can be manufactured to almost any length, restricted only by transportation to site.

Prior to lamination, the veneers are dried and the grains of each veneer are oriented in the same direction. This makes LVL stronger, straighter and more uniform than solid timber and overcomes some of timber's natural limitations such as strength-reducing knots. This gives orthotropic properties (different mechanical properties against different axes) in a similar way to the properties of sawn timber, rather than the isotropic properties (the same mechanical properties in each direction) in the plane of plywood. The added durability of being an engineered wood product means LVL is less prone to shrinking or warping. LVL can also support heavier loads and span longer distances than normal timber.

Section sizes are then cut from 1200 m wide sheets or "billets". The ability to cut different shapes from the LVL sheets allows for structural innovation using angular and curved shapes.

LVL provides a cost-effective and sustainable building material, delivering high structural reliability and strength.

Note: Some LVL members can be made with a few laminations laid up at right angles to enhance the shear strength of the member. These are known as Cross-Banded LVLs and may need to be specially ordered, as it is not a commonly stocked item.

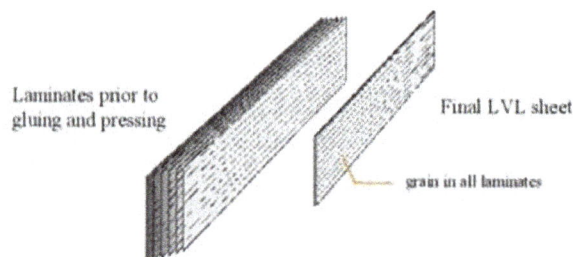

Structural composite lumber, including LVL, are a relatively recent innovation. They are the result of new technology and economic pressure to make use of new species and smaller trees that cannot be used to make solid sawn lumber. While plywood became widespread by the early twentieth century, the invention of LVL wasn't until the 1980s after the invention of oriented strand board. The National Design Specification for Wood Construction is generally updated on a 3- to 5-year cycle. The 1991 release is especially significant in that it is the first release which mentions LVL.

LVL is mentioned as a subcategory of structural glued laminated timber. The first explorations into engineered lumber happened during World War II in America. In 1942, an increased demand for wood caused a sudden timber shortage. The war industry utilized a panel material developed by the Homasote Company of Trenton, New Jersey made of wood pulp and ground newspaper to be used in place of siding and sheathing for buildings. The invention of laminated veneer lumber as known today can be attributed to Arthur Troutner. While glue laminated wood veneers were in use since the middle of the 19th century on a small scale for furniture and pianos, Troutner was the first to develop a laminated veneer lumber of a scale large enough to be used in construction. In 1971 "Micro=Lam LVL" was introduced. "Micro=Lam LVL" consisted of laminated veneer lumber billets 4 feet wide, 3-1/2 inches thick, and 80 feet long. Troutner proved the structural capabilities of his Micro=Lam product by building a house in Hagerman, Idaho using beams made of Micro=Lam. Most corporations considered Troutner's invention to be a niche product and it was not until the mid 1980s when logging became an environmental concern and corporations moved toward engineered lumber that LVL became widespread in use.

Qualities

Laminated veneer lumber is similar in appearance to plywood, although in plywood the veneers switch direction while stacking and in LVL the veneers all stack in the same direction. In LVL, the direction of the wood grain is always parallel to the length of the billet. The stacking of these veneers into a complete board, called a billet, creates a single piece of LVL sharing a common direction of wood grain. LVL is typically rated by the manufacturer for elastic modulus and allowable bending stress. Common elastic moduli are 12 GPa (1,700,000 psi); 13 GPa (1,900,000 psi); and 14 GPa (2,000,000 psi); and common allowable bending stress values are 19 MPa (2,800 psi); and 21 MPa (3,000 psi). Although the creation of LVL is often proprietary and thus its make-up is largely dependent on individual manufacturers, in general one cubic meter of North American Lumber is composed of 97.54% Wood, 2.41% of Phenol formaldehyde resin, 0.02% of Phenol resorcinol formaldehyde resin, and 0.03% fillers.

Manufacturing

LVL is commonly manufactured in North America by companies that also manufacture I-joists. LVL is manufactured to sizes compatible with the depth of I-joist framing members for use as beams and headers. Additionally, some manufacturers further cut LVL into sizes for use as chord-members on I-joists. In 2012, North American LVL manufacturers produced more than 1.2 million cubic metres (43.4 million cubic feet) of LVL in 18 different facilities, and in 2013 the production increased with more than 14%. It is not coincidental that LVL mills are often co-located with I-joist manufacturing facilities as many builders use a combination of I-joists and LVL in floor and roof assemblies.

Use

Because it is specifically sized to be compatible with I-joist floor framing, residential builders and building designers like the combination of I-joist and LVL floor and roof assemblies. LVL is considered to be a highly reliable building material that provides many of the same attributes associated with large sized timbers. LVL can also be used in combination with gluelam as an outer gluelam tension lam to increase the strength of the gluelam beam. However, due to the fact that the assembly adhesives limit the penetration of chemicals typically used to treat outdoor-rated lumber, LVL may not be suitable for outdoor load-bearing use. A deck built using pressure-treated LVL collapsed due to internal rotting of the twelve-year-old LVL components, although the LVL beams had passed regular visual inspections. The breakdown of LVL end uses in North America is 33% new single family residential construction, 25% residential renovations and upkeep, 8% new non-residential construction and 34% manufacturing furniture and other products.

Structural Composite Lumber

LVL belongs to the category of engineered wood called structural composite lumber. Other members of this category are parallel strand lumber (PSL) and laminated strand lumber (LSL). All members of this category are strong and predictable, and are thus interchangeable for some applications. PSL is made from veneers that are cut up into long strands and oriented parallel to its length before being compressed into its final shape. LSL is also made from strands rather than veneer, although the strands are shorter and aligned with less precision than PSL and is created as billets that are like a thick version of oriented strand board. Billets of PSL and LVL are very similar although their sizes are different. Billets of PSL can be as large as 12 inches wide and 60 feet long while LVL can range up to 4 feet wide and 80 feet long.

Architectural Roof Trusses

A truss is a structure comprising one or more triangular units. Each triangle is constructed with straight and usually slender members of timber, connected at the ends by joints. External loads, and the structure's reaction to those loads, act at the joints, resulting in forces that are either tensile or compressive.

The strength of a truss lies in its triangulation of banding members that work together to the advantage of the overall structure. For trusses, compression members often dictate the size of the elements, thus designs that have short compression members or restraint against lateral buckling are generally more efficient than trusses with longer compression members.

Within a building two forms of trusses can be found. Nail plated trusses are trusses hidden from view that use nail plates as connectors. Architectural trusses refer to those attractively detailed timber trusses, exposed to view. This guide focuses primarily on the application process of the latter.

The benefits of timber trusses are notable and numerous. Timber roof trusses are an ecologically sound choice, compared to conventionally pitched roofs, they use smaller dimension timbers that span greater distances and this in turn reduces the total timber volume contained within. Architectural timber trusses are lightweight, enabling speedy and efficient construction and installation that results that in a visual feature to be enjoyed for decades.

Framing

Lightweight timber construction typically comprises framed and braced structures to which one or more types of cladding are applied. Framing configurations can range from the closely spaced light timbers commonly seen in stud frame construction to large, more widely spaced timbers. A timber framed building can be placed on a concrete slab or on posts/poles or bearers resting on piers/stumps supported on pad footings.

Used in houses or multi-residential dwellings, lightweight timber construction offers the flexibility of a wide range of cost effective design options.

When the timber comes from sustainable sources, this construction method can be environmentally advantageous as it combines timber's low embodied energy with its capacity to store carbon.

Portal Frames

Timber portal frames are one of the most favoured structural applications for commercial and industrial buildings whose functions necessitate long spans and open interiors. As a material choice, timber offers designers simplicity, speed and economy in fabrication and erection.

Timber portal frames offer a strong, sound and superior structure. Structural action is achieved through rigid connections between column and rafter at the knees, and between the individual rafter members at the ridge. These rigid joints are generally constructed using nailed plywood gussets and on occasion, with steel gussets.

From material selection to finishing, this application guide provides a comprehensive overview of the process of using timber in the specification, fabrication and erection of portal frame structures.

Cross Laminated Timber

Cross-laminated timber (CLT) is a wood panel system that is gaining in popularity in the U.S. after being widely adopted in Europe. CLT is the basis of the tall wood movement, as the material's high strength, dimensional stability and rigidity allow it to be used in mid- and high-rise construction.

The Strength of CLT

CLT panels are made of layers of lumber boards (usually three, five or seven) stacked crosswise at 90-degree angles and glued into place. The panels can be manufactured at custom dimensions, though transportation restrictions dictate their length.

Applications for CLT include floors, walls and roofing. The panels' ability to resist high racking and compressive forces makes them especially cost-effective for multistory and long-span diaphragm applications.

Some specifiers view CLT as interchangeable with other wood products and building systems. Like other mass timber products, CLT can be used in hybrid applications with materials such as concrete and steel. It can also be used as a prefabricated building component, accelerating construction timelines.

Several factors contribute to a growing market for CLT and tall wood construction: advances in wood connectors, the development of hybrid materials and building systems, the commercialization of CLT and growth in off-site fabrication.

Usage of CLT

Alternating grains improve CLT panels' dimensional stability. The lumber boards typically vary in thickness from 5/8 inch to 2 inches and in width from 2.4 inches to 9.5 inches. Finger joints and structural adhesive connect the boards.

In structural systems, such as walls, floors and roofs, CLT panels serve as load-bearing elements. As such, in wall applications, the lumber used in the outer layers of a CLT panel is typically oriented vertically so its fibers run parallel to gravity loads, maximizing the wall's vertical load capacity. In floor and roof applications, the lumber used in the outer layers is oriented so its fibers are parallel to the direction of the span.

CLT's ability to resist high racking and compressive forces makes it a cost-effective solution for multistory and long-span diaphragm applications.

CLT's shear strength affords designers a host of new uses for wood. Those include wide prefabricated floor slabs, single-level walls and taller floor plate heights. As with other mass timber products, CLT can be left exposed in building interiors, offering additional aesthetic attributes.

Currently, U.S. building codes do not explicitly recognize mass timber systems, but this doesn't prohibit their use under alternative method provisions. The 2015 International Building Code (IBC) streamlines the acceptance of CLT buildings, recognizing CLT products when they are manufactured according to the Standard, ANSI/APA PRG-320: Standard for Performance Rated Cross Laminated Timber. In addition, CLT walls and floors may be permitted in all types of combustible construction, including Type IV buildings.

References

- Thomas Derdak, Jay P. Pederson (1999). International directory of company histories. 26. St. James Press. p. 82. ISBN 978-1-55862-385-9
- The-various-types-of-structural-steel-shapes: swantonweld.com, Retrieved 26 May 2018
- "ASTM A992?A992M Standard Specification for Structural Steel Shapes". American Society for Testing and Materials. 2006. doi:10.1520/A0992_A0992M-06A
- What-is-a-lally-column: lally-column.com, Retrieved 16 June 2018
- Naturally:wood Engineered wood Archived May 22, 2016, at the Portuguese Web Archive. Naturallywood.com. Retrieved on February 15, 2012
- Engineered-wood, glossary: kitchencabinetkings.com, Retrieved 31 March 2018
- Ridley-Ellis, Dan; Stapel, Peter; Baño, Vanesa (1 May 2016). "Strength grading of sawn timber in Europe: an explanation for engineers and researchers". European Journal of Wood and Wood Products. 74 (3): 291–306. doi:10.1007/s00107-016-1034-1
- Benefits-of-engineered-wood: reminetwork.com, Retrieved 15 May 2018
- Handbook of Steel Construction (9th ed.). Canadian Institute of Steel Construction. 2006. ISBN 978-0-88811-124-1

- What-is-glulam-applications-and-advantages-of-glulam-845106: thebalancesmb.com, Retrieved 25 April 2018

- "The wood from the trees: The use of timber in construction". Renewable and Sustainable Energy Reviews. 68: 333–359. 2017-02-01. doi:10.1016/j.rser.2016.09.107. ISSN 1364-0321

- Cross-laminated-timber-clt-handbook, products-and-systems: thinkwood.com, Retrieved 05 May 2018

- karenkoenig (2016-04-04). "Understanding & working with wood defects". Woodworking Network. Retrieved 2018-03-12

4

Bricks and Stones

A brick is a building material composed of clay that is used to make pavements and walls. It is of two kinds, fired and non-fired. Natural stone is used for construction. Large boulders of stone are excavated from bedrock, which are then cut and polished for creating border stones, tiles, stone blocks, etc. All the diverse topics such as cream city brick, London stock brick, Roman brick, Dutch brick, marble, limestone, etc. which are central to the use of bricks and stones in construction have been carefully examined in this chapter.

Brick

Bricks are used for building and pavement all throughout the world. In the USA, bricks were once used as a pavement material, and now it is more widely used as a decorative surface rather than a roadway material. Bricks are usually laid flat and are usually bonded forming a structure to increase its stability and strength. There are several types of bricks used many of them being about eight inches long and four inches thick.

Types of Bricks

Bricks are used as siding in the building industry due in part to its important characteristics and just because it can be a good affordable option. Below we summarize the benefits and applications of the most commonly used type of bricks.

Common Burnt Clay Bricks

Common burnt clay bricks are formed by pressing in molds. Then these bricks are dried and fired in a kiln. Common burnt clay bricks are used in general work with no special attractive appearances. When these bricks are used in walls, they require plastering or rendering.

Sand Lime Bricks

Sand lime bricks are made by mixing sand, fly ash and lime followed by a chemical process during wet mixing. The mix is then molded under pressure forming the brick. These bricks can offer advantages over clay bricks such as:

- Their color appearance is gray instead of the regular reddish color.
- Their shape is uniform and presents a smoother finish that doesn't require plastering.
- These bricks offer excellent strength as a load-bearing member.

Engineering Bricks

Engineering bricks are bricks manufactured at extremely high temperatures, forming a dense and strong brick, allowing the brick to limit strength and water absorption. Engineering bricks offer excellent load bearing capacity damp-proof characteristics and chemical resisting properties. These bricks are used in specific projects and they can cost more than regular or traditional bricks.

Concrete Bricks

Concrete bricks are made from solid concrete and are very common among homebuilders. Concrete bricks are usually placed in facades, fences, and provide an excellent aesthetic presence. These bricks can be manufactured to provide different colors as pigmented during its production.

Fly Ash Clay Bricks

Fly ash clay bricks are manufactured with clay and fly ash, at about 1,000 degrees C. Some studies have shown that these bricks tend to fail poor produce pop-outs, when bricks come into contact with moisture and water, causing the bricks to expand.

Advantages of Brick Construction

There are many advantages when bricks are used as part of the construction. The following list presents some of the most common advantages when using bricks instead of other construction materials.

- Aesthetic: Bricks offer natural and a variety of colors, including various textures.

- Strength: Bricks offer excellent high compressive strength.

- Porosity: The ability to release and absorb moisture is one of the most important and useful properties of bricks, regulating temperatures and humidity inside structures.

- Fire Protection: When prepared properly a brick structure can give a fire protection maximum rating of 6 hours.

- Sound Attenuation: The brick sound insulation is normally 45 decibels for a 4.5 inches brick thickness and 50 decibels for a nine-inch thick brick.

- Insulation: Bricks can exhibit above normal thermal insulation when compared to other building materials. Bricks can help regulate and maintain constant interior temperatures of a structure due to their ability to absorb and slowly release heat. This way bricks can produce significant energy savings, more than 30 percent of energy saving when compared to wood.

- Wear Resistant: A brick is so strong, that its composition provides excellent wear resistance.

- Efflorescence: Efflorescence forms on concrete structures and surfaces when soluble salts dissolved in water are deposited and accumulated on surfaces forming a visible scum.

- Durability: Brick is extremely durable and perhaps is the most durable man-made structural building material so far.

Other Brick Types

Bricks can also be classified depending on their geometry or shape. There are bricks in the form of bullnose, channel, coping, cownose and hollow bricks to name a few. Every one of these also has a unique function and characteristic that allows you to have the right brick at the right place.

Mudbrick

Mudbricks are pretty much what they sound like - bricks made of mud. Often, they are made on site from local soil, providing there is enough clay content. The soil is mixed with water and reinforcing materials such as straw and even cement and then pressed into wooden forms and allowed to set. The forms are removed and the bricks set aside to dry for up to several weeks. As they are made from natural materials they are a sustainable, recyclable, non-toxic and healthy form of building construction.

When building, mudbricks require suitable footings such as a course of regular fired bricks or concrete footings or slab. They are held together with a mud mortar mixture, which is similar in composition to the mudbrick mixture itself. Mudbricks, being soft and unfired, are easy to cut with hand tools, and can be formed into interesting shapes.

Once laying of the bricks is finished and the mortar is dry, the brick walls are often rendered with cement or mud-based renders, although many mudbrick walls are left au-naturale or are sealed with a transparent water-based sealant to improve weather resistance.

Mudbricks have many advantages, including low cost and low embodied energy (especially if they are made on site and not transported long distances) and ease of use. They also have high thermal mass (the ability to store and release heat), if the bricks are a minimum of 300 mm thick.

Earth buildings have excellent fire ratings, which make them suitable for building in bushfire prone areas and for the construction of fire rated walls within buildings. When designing new buildings consider using low pitched or curved roofs and parapet walls to reduce the impact of radiant heat and possibility of embers entering the roof area.

The drawbacks of mudbricks are, they are easily damaged, especially by rain and wind if not protected, and have high weight, making working with them a strenuous exercise if using larger bricks, which is quite common. They also are not good thermal insulators however insulation can be added to mudbrick walls with linings. However they are excellent sound insulators.

Cream City Brick

Cream City brick emerged from Milwaukee's kilns in the 1830s. The relatively high concentration of lime and sulfur in the clay found in the Menomonee Valley region created that creamy hue. Locally-produced, it became the least expensive "utility grade substance" of its day, making it a primary building block for churches, homes, factories and businesses for the next 70 years.

Milwaukee's position on railroads and its port on Lake Michigan allowed exportation of Cream City bricks for use in Great Lakes lighthouses -- including the charming one at Eagle Bluff in Door County's Peninsula State Park -- as well as commercial and industrial buildings in New York, Chicago and other American cities. They even made their way to Hamburg, Germany in 1859 on the M.S. Scott, noted as the second ship to leave Milwaukee for a foreign port. Architects overseas were interested, hence the Cream City brick that now graces a number of 19th-century buildings across Western Europe.

Milwaukee's pride in its brick showed in sports, too: a baseball club called the Milwaukee Cream City's compiled 15 wins against 45 losses in 1878, foreshadowing much of Milwaukee baseball for centuries to come. Despite a sixth-place finish, the attendance of the final game that season was 17,000. At that time, Cream City bricks were being created at the rate of over 15 million per year.

Cream City bricks brought a lighter color to buildings when they were new, but its porous nature and softer face meant dirt, smoke, weathering and pollutants could easily penetrate, and by the early 1900s many of the buildings constructed with Cream City brick lost their luster.

Meanwhile, changes in supply and cost structures, coupled with access to other building materials such as concrete, led to a decline in Cream City brick production shortly after 1900.

Restoration projects brought by developers, architects, preservationists, and others resurrected neighborhoods as well as an interest in the brick. Attempts at sandblasting buildings in the 1970s proved more harm than good - today, chemical composites restore the brick to its original look, as renovation projects across the city attest.

Cream City brick is "probably the toughest brick in America to clean because of its porosity," said Paul Jakubovich, a preservation planner in Milwaukee's Department of City Development.

Jakubovich notes that while Cream City bricks are "softer" than many other types, their toughness makes them among the most salvaged bricks for reconstruction. He points to a number of reconstruction projects in Arizona using Cream City bricks as an example. One can easily see from older buildings here in Milwaukee that the bricks hold up well.

Characteristics

Cream City bricks are made from red clay containing elevated amounts of lime and sulfur; this clay is common in regions of Wisconsin, especially near Milwaukee. When the bricks are fired, they become creamy-yellow in color.

Cottage in Racine, Wisconsin.

Although light-colored when first made, Cream City bricks are porous, causing them to absorb dirt and other pollutants; this tends to make structures constructed out of them dark-colored as time passes. Once Cream City bricks absorb pollutants, they are difficult to clean, a problem, which restoration experts in Milwaukee have been facing since the 1970s. Initially, sandblasting was attempted; however, it not only proved to be ineffective, but damaged the bricks. Currently, chemical washes are accepted as the most effective method of cleaning Cream City bricks. The historic Trimborn Farmhouse in Greendale, Wisconsin is an example of brick that has been cleaned to reveal its original color.

Structures Built with Cream City Brick

Trinity Evangelical Lutheran Church, located in Milwaukee, Wisconsin, is an example of a building constructed with Cream City brick, though its cream color has been darkened by the elements.

Cream City bricks are well known for their durability; many buildings constructed with them in the 19th century still stand today. An example of the durability of Cream City brick is Trinity Evangelical Lutheran Church, which was built more than 125 years ago. However, since there were numerous brick makers in the area, brick quality varied and some of the bricks were not manufactured properly: the Big Sable Point Lighthouse was constructed of Cream City brick, but it had degraded so much in 35 years that it had to be encased in iron plating.

Because the regional headquarters of the United States Lighthouse Board responsible for building lighthouses around Lake Michigan was located in Milwaukee, many of them are built with Cream City bricks, including Kenosha Light, the Eagle Bluff Lighthouse, the McGulpin Point Light, the Old Mackinac Point Light, and many others.

Cream City bricks were widely exported, making their way not only to American cities such as Chicago and New York, but also to Western Europe, including Hamburg, Germany.

London Stock Brick

Often referred to as the 'London Stock' for its common usage in the British capital and surrounding areas, the Yellow Stock brick tells the story of nineteenth and twentieth-century London in style. With their authentic worn, but no less distinctive shades of yellow colouring, they remind us of the chipper, cheerful and sunny British spirit that endures despite the cold, wet climate. The iconic yellow colouring comes from the variety of minerals in the soft, dense clay of the Thames, which come to life in an assortment of yellows when fired into bricks.

Originating in the late Georgian era, these London Stock bricks built many of the houses and structures that stood at defining moments in British history. Yellow Stocks occasionally sport a shock of black flecks that come from ash in the clay going black when fired, but to us, these dark marks symbolise something more. For us, they tell a tale of the sooty industrial revolution, the Victorian sprawling urbanisation of the capital, and the ashes and flames of the 1940s when London survived the Blitz. Like the people of the city, these bricks and the buildings they were part of stood firm, and many still stand today.

Bricks were also made in clay areas surrounding London. During the 19th century and early 20th century 5 million yellowish stock bricks a year were supplied from the brickfields of Yiewsley and Starveall Middlesex for the building of the `new` London. Bricks were also made in Kent, Essex and other areas where they could be imported to London by rail. In Stock, Essex there is a common belief that "Stock Bricks" originated there; bricks were certainly made there, but the name is a coincidence, stock being a common English word with many meanings and also a common place-name element.

Air pollution in London during the 19th century and early 20th century commonly caused the bricks to receive a sooty deposition over time, turning the bricks greyish or even black, but the removal of contaminants from the air following the passing of the Clean Air Act in 1956, has enabled older buildings to be cleaned and new buildings to retain their natural colour.

In the 19th century, London stock bricks were available in a variety of grades priced according to their consistency and their regularity of shape and colour. High-grade bricks were used for face work and lower grades were bought for use as internal bricks. Unfortunately it seems to have been common practice for a high-grade brick to be broken in half so that it could be used twice, each end appearing as a header in the wall. The result of this parsimony was that the wall was deficient in bonding bricks, i.e. bricks tying the outer skin of brickwork back to the inner part of the wall, often resulting in the outer skin peeling away from the inner and bulging out. This issue, known as snapped or snap headers, leads to walls, which need to be repaired either by rebuilding or by fitting various types of proprietary tie.

Stock Bricks

The term 'stock brick' can either indicate the common type of brick stocked in a locality, or a hand-made brick made using a stock. A stock or stock board is an iron-faced block of wood fixed to the surface of the moulder's bench. The brick mould fits over the stock; the brick maker fills the mould with prepared clay and cuts it off with a wire level with the top of the mould, before turning out the 'green' brick onto a wooden board called a pallet for drying and firing.

Reclaimed London stock bricks are sought after for decorative and conservation use. The mortar usually used with them in original construction was lime mortar which is much softer and weaker than modern cement based mortar and can be cleaned off second hand bricks easily leaving them ready for re-use. Nevertheless, the supply of second-hand stocks cannot always meet the demand, and 'new' second-hand stocks can be obtained from builders' merchants. Unfortunately, some of the new products are painted white or black to simulate the whitewashed or soot blackened surfaces often encountered in the real second hand bricks - making them unsuitable for face-work.

Mortar

Most London stock bricks are more or less porous, as is the lime mortar in which they have tradition-ally been laid. The pointing should be flush pointing so that rain water can run down off the surface and not be encouraged to soak into the wall as is the case with recessed or struck pointing. When used in this way the brickwork does not get wet all the way through and is thus effectively waterproof.

Lime mortar tends to weaken in London's acidic rainwater and needs repointing several times a century. It has been common since the widespread availability of Portland cement to see London stock brickwork repointed using much stronger cement mortar. As repointing consists of replacing the outer 20 - 40mm of mortar, the effect of this is to make the outer 20 - 40 mm of the brickwork harder and stronger than the interior of the wall. This can lead to spalling of the brick surface, and can also encourage the bulging associated with snapped headers.

Staffordshire Blue Brick

Staffordshire blue brick is a strong type of construction brick, originally made in Staffordshire, England.

The brick is made from the local red clay, Etruria marl, which when fired at a high temperature in a low-oxygen reducing atmosphere takes on a deep blue colour and attains a very hard, impervious surface with high crushing strength and low water absorption.

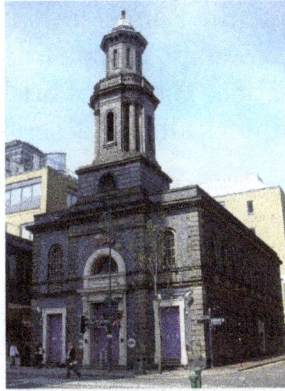
Staffordshire blue brick, used here for its appearance rather than its high strength

Brickworks were a key industry across the whole Black Country throughout the 19th and 20th centuries, and were considered so important that they were designated as a reserved occupation during World War Two. The Black Country was a major producer of clay for brickmaking, often mined from beneath the famous 30 foot Staffordshire coal seam. This is an ancient industry, and probably took place from the 17th century at least, however the industry really took off in the 19th century. A key date is 1851 when the Joseph Hamblett brickworks was founded in West Bromwich, which became one of the largest producers of the famous Staffordshire blue bricks. Other sites produced these too, including Albion in West Bromwich, Cakemore works at Blackheath, Springfield at Rowley Regis, John Sadler, Blades and New Century at Oldbury, Coneygre at Tipton and Bentley Hall near Darlaston.

Viaduct carrying the line and platforms of Birmingham Snow Hill station

This type of brick was used for foundations and was extensively used for bridges and tunnels in canal construction, and later, for railways. Its lack of porosity makes it suitable for capping brick walls, and its hardwearing properties makes it ideal for steps and pathways. It is also used as a general facing brick for decorative reasons. Staffordshire Blue bricks have traditionally been "Class A" with a water absorption of less than 4.5%.

Roman Brick

By the 2nd century AD, brick-making was so woven into the fabric of Roman life that Plutarch warned:

"No man ever wetted clay and then left it, as if there would be bricks by chance and fortune."

The Romans made fired clay bricks, and the Roman legions, which operated mobile kilns, introduced bricks to the whole of the Roman Empire. Roman bricks are often stamped with the mark of the legion that supervised their production. The use of bricks in southern and western Germany, for example, can be traced back to traditions already described by the Roman architect Vitruvius.

Preferring to make their bricks in the spring, the Romans held on to their bricks for 2 years before they were used or sold. They only used clay, which was whitish or red for their bricks.

Close-up view of the wall of the Roman shore fort at Burgh Castle,
Norfolk, showing alternating courses of flint and brickwork.

The kiln fired bricks were generally 1 or 2 Roman feet by 1 Roman foot, but with some larger bricks at up to 3 Roman feet. The Romans preferred this type of brick making during the first century of their civilisation and used the bricks for public and private buildings all over the empire.

The Constantine Basilica in Trier utilises Roman brick.

Roman brick was almost invariably of a lesser height than modern brick, but was made in a variety of different shapes and sizes. Shapes included square, rectangular, triangular and round, and the largest bricks found have measured over three feet in length. Ancient Roman bricks had a general size of 1½ Roman feet by 1 Roman foot, but common variations up to 15 inches existed. Other brick sizes in Ancient Rome included 24" x 12" x 4", and 15" x 8" x 10". Ancient Roman bricks found in France measured 8" x 8" x 3". The Constantine Basilica in Trier is constructed from Roman bricks 15" square by 1½" thick. There is often little obvious difference (particularly when only fragments survive) between Roman bricks used for walls on the one hand, and tiles used for roofing or flooring on the other, and so archaeologists sometimes prefer to employ the generic term Ceramic Building Material (or CBM).

The Romans perfected brick making during the first century of their empire and used it ubiquitously, in public and private construction alike. The Romans took their brickmaking skills everywhere they went, introducing the craft to the local populations. In the British Isles, the introduction of Roman brick by the Ancient Romans was followed by a 600 - 700 year gap in major brick production.

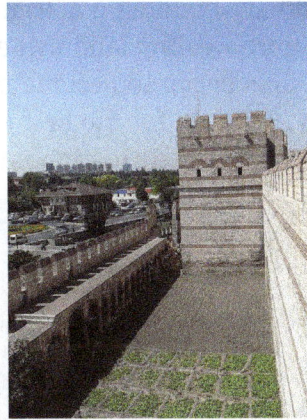

The city walls of Constantinople, showing several courses of brickwork.

When building in masonry, the Romans often interspersed the stonework at set intervals with thin courses of bricks, sometimes known as "bonding tiles." This practice gave the structure added stability. It also had a secondary aesthetic effect of creating a polychromatic appearance.

Close-up view of the wall of the Roman shore fort at Burgh Castle, Norfolk, showing alternating courses of flint and brickwork.

Ancient Roman Brick Stamps

Ancient Roman stamp on a hypocaust brick, used by the third cohort of Roman citizens from Thrace.

Around the middle of the 1st century BC Roman brick makers began using unique identifying stamps on their bricks. This was done because the brick factories were responsible for their quality and so they could be held responsible for defective products. The first of these brick stamps were simple and included minimal information such as, the name of a person and sometimes the name of the brickyard the brick was produced in. These earliest Roman brick stamps were emblazoned into the wet clay using a hardwood or metal mold prior to the firing of the brick. As the early Roman Empire progressed fired brick became the primary building material and the number of brick producers increased dramatically as more and more wealthy land owners began to exploit clay deposits on their land for brick-making. Brick stamps began to become more complex and include more and more information. In 110, the stamps included, for the first time, the name of the consuls for the year of production, which allows modern observers to pinpoint the year a brick was created.

Brickstamp with Latin stamped impression inscription: "

These brick stamps, once viewed more as a curiosity than archaeological artifacts, allow scholars to learn about the demand for bricks in Ancient Rome because through the dates on the stamps they provide a chronology. Today, brick stamp discoveries are carefully documented and that documentation, combined with the use of architectural context, has helped provide a reliable method of dating Ancient Roman construction. In addition, brick stamps have proved helpful in determining general Ancient Roman chronology.

This brick, found in Ostia Antica, is stamped with
the Consular names Paetinus and Apronianus.

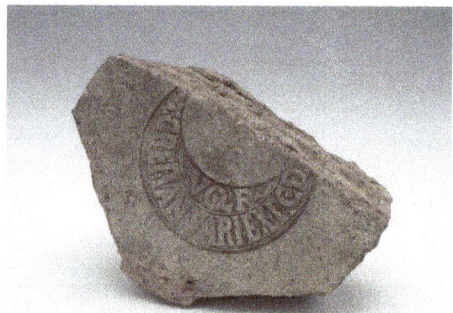

From the Harvard Art Museum.

2nd Century Imperial Brick used to build the Basilica di Santa Pudenziana.

Roman Bricks and Styles

One of the things the Romans are most famous for is their architecture. The Romans brought a lot of new ideas to architecture, of which the three most important are the arch, the baked brick, and the use of cement and concrete. Romans built the first major public bath building in Rome and also the Hadrian"s Pantheon, a temple to all the gods, which used brick and concrete to build a huge dome.

The Romans developed a very effective kind of mortar by mixing pozzolana, a volcanic ash of the region around Naples, with lime; they obtained cement, which was resistant to water. In his work De Architectura (a treatise on architecture dedicated to the Emperor Augustus) Vitruvius so described pozzolana:

There is a species of sand which, naturally, possesses extraordinary qualities. It is found about Baiæ and the territory in the neighbourhood of Mount Vesuvius; if mixed with lime and rubble, it hardens as well under water as in ordinary buildings. This seems to arise from the hotness of the earth under these mountains, and the abundance of springs under their bases, which are heated either with sulphur, bitumen, or alum, and indicate very intense fire. The inward fire and heat of the flame which escapes and burns through the chinks, makes this earth light; the sand-stone (tophus), therefore, which is gathered in the neighbourhood, is dry and free from moisture. Since, then, three circumstances of a similar nature, arising from the intensity of the fire, combine in one mixture, as soon as moisture supervenes, they cohere and quickly harden through dampness; so that neither the waves nor the force of the water can disunite them.

Roman bricks (left) Roman bricks at S. Saba; (centre) walls supporting
the Imperial Palaces; (right) brick decoration at Anfiteatro Castrense.

Most Roman buildings are made up of opus caementicium, a sort of concrete which was laid into timber structures until it hardened. The resulting walls were very solid, but not nice to see, so very often some sort of facing was applied.

Brickworks Brickwork (opus latericium) at Pantheon (left) and at Palazzo Madama (right).

The Romans made use of fired bricks; the manufacturing of bricks was perfected during the first century AD: factories branded their products, as they were responsible for their quality; bricks were generally longer and narrower than the bricks we use today. There were also round and triangular bricks, which were used to imitate columns and other architectural motifs.

The way the bricks were laid is often associated with a specific period: the texture of course-laid brickwork at the time of Emperor Hadrian impressed Renaissance architects who imitated it in many buildings. Contemporary architects have used "Roman brick" too.

"Opus reticulatum" at Villa Adriana (left) and near S. Saba (right).

An unusual kind of facing was based on specially-shaped tufa stones: the points of the stones were inserted into opus caementicium, while their square bases formed a diagonal pattern resembling a fishing net (reticulatum). Today many streets of Rome are paved with porphyry stones, which create a similar diagonal pattern.

Opus listatum (left) "Opus listatum" at Porta
S. Paolo; (right) "Opus mixtum" at Delphi.

Bricks were relatively expensive and their laying required a lot of manpower: for these reasons in the last centuries of the Roman Empire new facing patterns were developed to reduce the use of bricks; they were replaced by tufa pieces or by other materials including pieces of marble and other stones coming from ruined buildings. This resulted in a stripe design which characterized the buildings of the Late Empire as showed by the walls of Constantinople. These stripes passed on to Byzantine Art; they then influenced the design of Muslim mosques to finally return to Italy, where many medieval cathedrals (e.g. Orvieto) were decorated with black and white stripes.

Dutch Brick

Dutch brick (in dutch IJsselsteen) is a small type of yellow brick made in the Netherlands, or similar brick, and an architectural style of building with brick developed by the Dutch. The brick, made from clay dug from riverbanks or dredged from riverbeds of the river Ijssel and fired over a long period of time, was known for its durability and appearance.

Close-up of Dutch bricks with inscription

Traditional Dutch brick architecture is characterized by rounded or stepped gables. The brick was imported as ballast into Great Britain and the colonies in the east of America. Trinity College, Dublin, Ireland, founded in 1591, was originally built of red Dutch brick. Dutch brickmakers emigrated to New Netherland in America, where they built kilns for firing bricks locally. Bricks were being burned in New Amsterdam (New York) by 1628, but the imported bricks were of better quality. At first the bricks were used only for chimneys, but they were later used to face the lower story of the house, and then the entire house. Most of the surviving "Dutch Colonial" houses in New York do not in fact follow Dutch architectural practices, but there are several examples in Albany County, which do.

Bricks were also exported by the Dutch for major buildings in their colonies in the east and around the world. The Castle of Good Hope in Cape Town, South Africa, was built in 1666, and its entrance was made of the small yellow bricks called *ijselstene* (IJssel stones). Christ Church in Malacca, Malaysia, the oldest Dutch church building outside the Netherlands, was made of Dutch bricks that had been brought as ballast in ships from the Netherlands, coated with Chinese plaster.

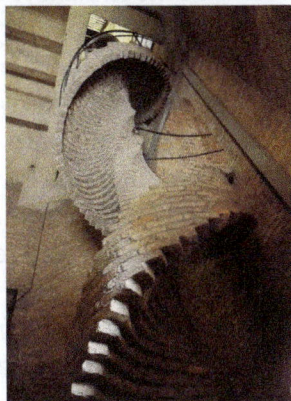

Spiral staircase to the carillon of the Dutch Reformed Church of IJsselstein

Background and Manufacture

The word "brick" may be of Dutch origin. A 1901 dictionary of architecture defines "brick" as "a regularly shaped piece of clay hardened in the sun or by the heat of a kiln and intended for building; commonly one of very many pieces of uniform size." The "Dutch Brick" is described as "a hard, light-coloured brick originally made in Holland and used in England for pavements; hence a similar brick made in England."

Until well into the twentieth century the manufacture of brick in the Netherlands (and elsewhere) used manual labor mostly, and the low-paid workers involved in the industry were as marginal socially as the manufacturing industry geographically—the raw materials were gathered on river banks, and the firing of the bricks took place well away from towns and farms to lessen any nuisance caused by fire and smoke. Workers, as was noted in municipal reports, often already belonged to the lower levels of society and were frequently simply let go at the end of the season, adding to the municipal burdens: "As the number of brick kilns increases, so does poverty," according to the 1873 report of the Ubbergen municipality, near Nijmegen, on the Waal river.

Manual brick manufacture.

The clay for the bricks was dug from river banks (of the Waal, Rhine, and IJssel rivers) and other open-air locations, and was left outside (in a mound called the *kleibult*) through the winter so that any organic material could decay; the weather (rain, frost, drought) helped make the clay more manageable. At the end of this period the clay was mixed with sand and other materials, a process done by foot, by workers stomping on the clay. It was then molded into the proper shape by an artisan, the *tichelaar* ("brickmaker"). Children handed the brickmaker the raw material and removed the shaped bricks. Child labor was common in the industry: until well into the nineteenth century children eight years old and younger worked 16 hours per day, and children four years old stacked and carried bricks for hours at a stretch. Molds were moistened with water and strewn with sand to enable the shaped brick to be more easily removed. The "raw" or "green" stones were laid out in long rows to dry and when they were dry enough they were stood up on their side so the bottom could dry; this work was often done by women and children. Often it was the women who did the much heavier labor of moving the dried bricks to the ovens, hauling wheelbarrows with loads of up to 80 kilograms, and stacking and preparing the ovens and tending to the fire (which burned peat or coal). Ovens came in two types—a single-use construction of the kind used in the production of charcoal, and a more permanent type, basically consisting of two walls one meter and a half thick. Ovens could hold up to a million bricks. Masonry bricks were fired between 900° C (1,650° F) and 1,125° C (2,057° F), klinkers between 1,150° C (2,100° F) and 1,250° C (2,280° F). Typically, bricks were baked at low heat for two weeks to remove all remaining moisture from the clay, and then for four weeks at a higher temperature, followed by two weeks of cooling down.

Since the klinker was partially vitrified by being fired at a higher temperature it was harder than the standard. Klinkers were imported into England for use as paving.

Small, yellow Dutch bricks used to be imported into the United States, and as of 1840 there were still old buildings in New York faced in these bricks. They were considered superior in appearance and in durability. An 1888 report noted that "in New York and other Atlantic cities we find houses built of brick brought from Holland fully two hundred years ago, without a flaw or sign of decay, and apparently as firm and sound as when first laid in the wall."

Europe

Old Dutch farmhouse with thatched roof

Houses found today in Zeeland are closer in appearance to the fine Dutch brick houses of New York than are houses from other parts of the Netherlands. Brick farm houses built separately from barns are found in Zeeland, but none have survived in other locations. Unlike the common practice in New York, the farm houses in Zeeland do not have separate outside doors for each room. The Dutch also used bricks to pave the roads, or chaussees, in the Netherlands.

By the 1640s the Dutch were considered to be the leaders in Europe both in making bricks and in bricklaying. The Summer Garden in Saint Petersburg, Russia, exhibits the work of Dutch brickmakers and bricklayers. Saint Michael's Castle, built in Saint Petersburg between 1797 and 1801 for the Emperor Paul I, is "an enormous quadrangular pile, of red Dutch brisk, rising from a massy basement of hewn granite." Sans Souci, the palace built for Frederick the Great in Potsdam, was built with a facade of rich red Dutch brick.

In recent years the Dutch brick industry has attracted unwelcome attention from the European Union (EU) competition authorities. In the early 1990s the industry had excess capacity due to technological advances, competition from other materials and an economic slowdown. Producers with combined market share of 90% agreed to reduce capacity, shutting down the older and inefficient plants. The producers compensated those who closed plants. However, the agreement also included fixing production quotas and fining members who produced more than their quota. The members of what was in effect a cartel were forced to drop the quota agreement by the EU.

"Dutch Houses", Topsham, Devon, England

Great Britain and Ireland

Imported Dutch brick was often used in buildings in England in the 17th and 18th centuries. In Dartmouth, a house built in 1664 for mariner Robert Plumleigh had traditional timber-framed architecture but included elaborate star-shaped chimney stacks made from imported Dutch brick. Houses in Topsham, Devon, also used Dutch brick for chimneys, window heads and dressing. One house from the late 17th century in Dutch Court in Topsham is built entirely of Dutch brick. The ports of Exeter and Topsham both shipped wool to the Netherlands, and the returning ships brought bricks as ballast from Amsterdam or Rotterdam.

Trinity College, Dublin, Ireland, founded in 1591, was originally built of red Dutch brick. Jigginstown House in Naas, County Kildare, Ireland, was built by John Allen for Thomas Wentworth, 1st Earl of Strafford using Dutch brick "of the most superior manufacture". The Red House in Youghal, Ireland, was built of red Dutch brick in 1710 by the Dutch architect Leuventhal for the Uniacke family.

United States

Van Alen House, Kinderhook, New York, built around 1737

In general, bricks were not imported to the American colonies. Probably none were imported to Virginia and Maryland, but in New England there was one possible example in New Haven, and there are records documenting the shipment of 10,000 bricks to Massachusetts Bay in 1628 and several thousand bricks being shipped to New Sweden. It is possible that the terms "Dutch brick" and "English brick" referred to the size of the locally made bricks, with the Dutch bricks being the smaller. However, in New Netherland there are records of brick being imported from the Netherlands as ballast in 1633, and of continued shipments until the American Revolution. Bricks were being burned in New Amsterdam (New York) by 1628, but the imported bricks were of better quality. At first, the bricks were used only for chimneys, but they were later used to face the lower story of the house, and then the entire house.

Dutch brick makers emigrated to New Netherland, where they built kilns for firing bricks locally. In New Amsterdam, brick was used for the director general's house, the counting house, the city tavern and other important buildings. Houses were gable-ended, often with stepped designs, and the bricks ranged in color from yellow or red to blue or black. An account of New York published in 1685 said, "The town is broad, built with Dutch brick, consisting of above five hundred houses, and the meanest not valued under an hundred pounds." A New Englander who visited New York in 1704, forty years after the Dutch had yielded the town to the British, admired the appearance of the glazed brickwork of the houses of "diverse coullers and laid in Checkers". In 1845 there was still a one-story Dutch brick house built in 1696 in Flatbush, Brooklyn. The date and the owner's initials were formed by blue and red glazed bricks.

General Wayne Hotel, Philadelphia, USA, c. 1803, gambrel
roof added and enlarged 1866. Dutch Colonial style

A view of part of Albany, New York, as it was in 1814 shows a mixture of Dutch, English and Federal styles, although Dutch brick was reportedly used for one of the English-style houses. One house in the Dutch style was said to date from the American Revolution. If so, it would have been one of the last genuine Dutch-style houses to be built in the United States, reflecting the conservative Dutch culture of Albany at that time.

Most of the surviving "Dutch Colonial" houses in New York do not in fact follow Dutch architectural practices, but there are seven in Albany County that do. The houses have a wood frame with brick walls as a decorative shell. They each have two parapet gables edged with "mouse toothing" ornamental brickwork. All the Dutch brick buildings used iron wall anchors spread across several bricks to tie the brick shell to the wooden frame of the house. Sometimes the anchor gives the date of construction. The brickwork of the houses incorporated various designs including spear shapes and a form like a fleur-de-lis.

Other Dutch Colonies

Fort Zeelandia (Taiwan). Original wall of imported red bricks
laid by the soldiers of the Dutch East India Company.

Dutch bricks and brickwork were also imported and utilized in other colonies throughout the Dutch Empire in Asia, Africa, and the Caribbean. Fort Zeelandia was built on a small island off Tainan in Formosa (Taiwan) between 1624 and 1634 after the Dutch acquired Formosa from China as a trading colony. It was built using bricks from Batavia (Jakarta), where the Dutch East India Company had its headquarters. After a siege in 1662, the Dutch surrendered the fort to Koxinga, a Ming dynasty general. The fort was destroyed by an explosion in 1873 when a shell from a British warship blew up the ammunition storehouse. The masonry was later used for other purposes. All that remains is part of the southern wall.

The Castle of Good Hope in Cape Town, South Africa, was built in 1666. The gateway was built in 1682, with a pediment and two pilasters of grey-blue stone, and an entrance made of the small yellow bricks called *ijseltene* (IJssel stones).

Christ Church, Malacca, Malaysia, is the oldest Dutch church building outside the Netherlands. It was built by the local Dutch burghers after the town had been taken from the Portuguese, and was completed in 1753. The church covers 82 by 41 feet (25 by 12 m), with a ceiling 40 feet (12 m) high. The foundations were local laterite blocks. The walls, which are massive, were made of Dutch bricks that had been brought as ballast in ships from the Netherlands, and they were coated with Chinese plaster.

On the island of Sint Eustatius in the Netherlands Antilles, the houses were built from local volcanic stone, from imported wood, or from red or yellow Dutch brick imported from the Netherlands. The traditional masonry houses were both large and solid. The country house of Johannes de Graaff, who commanded Sint Eustatius from 1776 to 1781, features a 33.6-by-9.7-foot (10.2 by 3.0 m) duck pond made of brick.

Marble

Marble is a metamorphic rock that forms when limestone is subjected to the heat and pressure of metamorphism. It is composed primarily of the mineral calcite ($CaCO_3$) and usually contains other minerals, such as clay minerals, micas, quartz, pyrite, iron oxides, and graphite. Under the conditions of metamorphism, the calcite in the limestone recrystallizes to form a rock that is a mass of interlocking calcite crystals. A related rock, dolomitic marble, is produced when dolostone is subjected to heat and pressure.

Pink Marble: A piece of pink marble about four inches (ten centimeters) across. The pink color is most likely derived from iron.

Sources of Marble

Most marble forms at convergent plate boundaries where large areas of Earth's crust are exposed to regional metamorphism. Some marble also forms by contact metamorphism when a hot magma body heats adjacent limestone or dolostone.

Before metamorphism, the calcite in the limestone is often in the form of lithified fossil material and biological debris. During metamorphism, this calcite recrystallizes and the texture of the rock changes. In the early stages of the limestone-to-marble transformation, the calcite crystals in the rock are very small. In a freshly broken hand specimen, they might only be recognized as a sugary sparkle of light reflecting from their tiny cleavage faces when the rock is played in the light.

As metamorphism progresses, the crystals grow larger and become easily recognizable as interlocking crystals of calcite. Recrystallization obscures the original fossils and sedimentary structures of the limestone. It also occurs without forming foliation, which normally is found in rocks that are altered by the directed pressure of a convergent plate boundary.

Recrystallization is what marks the separation between limestone and marble. Marble that has been exposed to low levels of metamorphism will have very small calcite crystals. The crystals become larger as the level of metamorphism progresses. Clay minerals within the marble will alter to micas and more complex silicate structures as the level of metamorphism increases.

Ruby in Marble: Marble is often the host rock for corundum, spinel, and other gem minerals. This specimen is a piece of white marble with a large red ruby crystal from Afghanistan. Specimen is about 1 1/4 inches across (about 3 cm).

Physical Properties and Uses of Marble

Marble occurs in large deposits that can be hundreds of feet thick and geographically extensive. This allows it to be economically mined on a large scale, with some mines and quarries producing millions of tons per year.

Most marble is made into either crushed stone or dimension stone. Crushed stone is used as an aggregate in highways, railroad beds, building foundations, and other types of construction. Dimension stone is produced by sawing marble into pieces of specific dimensions. These are used in monuments, buildings, sculptures, paving and other projects.

Color

Marble is usually a light-colored rock. When it is formed from a limestone with very few impurities, it will be white in color. Marble that contains impurities such as clay minerals, iron oxides, or bituminous material can be bluish, gray, pink, yellow, or black in color.

Marble of extremely high purity with a bright white color is very useful. It is often mined, crushed to a powder, and then processed to remove as many impurities as possible. The resulting product is called "whiting." This powder is used as a coloring agent and filler in paint, whitewash, putty, plastic, grout, cosmetics, paper, and other manufactured products.

Gray Marble: This specimen has calcite cleavage faces up to several millimeters in size that are reflecting light. The specimen is about twoinches (5 cm) across.

Acid Reaction

Being composed of calcium carbonate, marble will react in contact with many acids, neutralizing the acid. It is one of the most effective acid neutralization materials. Marble is often crushed and used for acid neutralization in streams, lakes, and soils.

It is used for acid neutralization in the chemical industry as well. Pharmaceutical antacid medicines such as "Tums" contain calcium carbonate, which is sometimes made from powdered marble. These medicines are helpful to people who suffer from acid reflux or acid indigestion. Powdered marble is used as inert filler in other pills.

Calcium carbonate medicines: Marble is composed of calcium carbonate. That makes it very effective at neutralizing acids.

Highest purity marble is often crushed to a powder, processed to remove impurities, and then used to make products such as Tums and Alka-Seltzer that are used for the treatment of acid indigestion.

Crushed marble is also used to reduce the acid content of soils, the acid levels of streams, and as an acid-neutralizing material in the chemical industry.

Hardness

Being composed of calcite, marble has a hardness of three on the Mohs hardness scale. As a result, marble is easy to carve, and that makes it useful for producing sculptures and ornamental objects. The translucence of marble makes it especially attractive for many types of sculptures.

The low hardness and solubility of marble allows it to be used as a calcium additive in animal feeds. Calcium additives are especially important for dairy cows and egg-producing chickens. It is also used as a low-hardness abrasive for scrubbing bathroom and kitchen fixtures.

Ability to Accept a Polish

After being sanded with progressively finer abrasives, marble can be polished to a high luster. This allows attractive pieces of marble to be cut, polished, and used as floor tiles, architectural panels, facing stone, window sills, stair treads, columns, and many other pieces of decorative stone.

Limestone

Limestone is a sedimentary rock, composed mainly of skeletal fragments of marine organisms such as coral, forams and molluscs. Its major materials are the minerals calcite and aragonite, which are different crystal forms of calcium carbonate ($CaCO_3$).

About 10% of sedimentary rocks are limestones. The solubility of limestone in water and weak acid solutions leads to karst landscapes, in which water erodes the limestones over thousands to millions of years. Most cave systems are through limestone bedrock.

Limestones has numerous uses: as a building material, an essential component of concrete (Portland cement), as aggregate for the base of roads, as white pigment or filler in products such as toothpaste or paints, as a chemical feedstock for the production of lime, as a soil conditioner, or as a popular decorative addition to rock gardens.

The first geologist to distinguish limestone from dolomite was Belsazar Hacquet in 1778.

Like most other sedimentary rocks, most limestones are composed of grains. Most grains in limestones are skeletal fragments of marine organisms such as coral or foraminifera. Other carbonate grains comprising limestones are ooids, peloids, intraclasts, and extraclasts. These organisms secrete shells made of aragonite or calcite, and leave these shells behind when they die.

Limestone often contains variable amounts of silica in the form of chert (chalcedony, flint, jasper, etc.) or siliceous skeletal fragment (sponge spicules, diatoms, radiolarians), and varying amounts of clay, silt and sand (terrestrial detritus) carried in by rivers.

Some limestones do not consist of grains at all, and are formed completely by the chemical precipitation of calcite or aragonite, i.e. travertine. Secondary calcite may be deposited by supersaturated meteoric waters (groundwater that precipitates the material in caves). This produces speleothems, such as stalagmites and stalactites. Another form taken by calcite is oolitic limestone, which can be recognized by its granular (oolite) appearance.

The primary source of the calcite in limestones is most commonly marine organisms. Some of these organisms can construct mounds of rock known as reefs, building upon past generations. Below about 3,000 meters, water pressure and temperature conditions cause the dissolution of calcite to increase nonlinearly, so limestone typically does not form in deeper waters. Limestones may also form in lacustrine and evaporite depositional environments.

Calcite can be dissolved or precipitated by groundwater, depending on several factors, including the water temperature, pH, and dissolved ion concentrations. Calcite exhibits an unusual characteristic called retrograde solubility, in which it becomes less soluble in water as the temperature increases.

Impurities (such as clay, sand, organic remains, iron oxide, and other materials) will cause limestones to exhibit different colors, especially with weathered surfaces.

Limestone may be crystalline, clastic, granular, or massive, depending on the method of formation. Crystals of calcite, quartz, dolomite or barite may line small cavities in the rock. When conditions are right for precipitation, calcite forms mineral coatings that cement the existing rock grains together, or it can fill fractures.

Travertine is a banded, compact variety of limestone formed along streams; particularly where there are waterfalls and around hot or cold springs. Calcium carbonate is deposited where evaporation of the water leaves a solution supersaturated with the chemical constituents of calcite. Tufa, a porous or cellular variety of travertine, is found near waterfalls. Coquina is a poorly consolidated limestone composed of pieces of coral or shells.

During regional metamorphism that occurs during the mountain building process (orogeny), limestones recrystallize into marble.

Limestones are a parent material of Mollisol soil group.

Uses

The Megalithic Temples of Maltasuch as Hagar Qim are built entirely of limestones. They are among the oldest free-standing structures in existence.

Limestones are very common in architecture, especially in Europe and North America. Many landmarks across the world, including the Great Pyramid and its associated complex in Giza, Egypt, were made of limestone. So many buildings in Kingston, Ontario, Canada were, and continue to be, constructed from it that it is nicknamed the 'Limestone City'. On the island of Malta, a variety of limestones called Globigerina limestones was, for a long time, the only building material available, and is still very frequently used on all types of buildings and sculptures. Limestones is readily available and relatively easy to cut into blocks or more elaborate carving. It is also long-lasting and stands up well to exposure. However, it is a very heavy material, making it impractical for tall buildings, and relatively expensive as a building material.

The Great Pyramid of Giza, one of the Seven Wonders of the Ancient World had an outside cover made entirely from limestone.

Riley County Courthouse in Manhattan, Kansas, USA is also built of limestone

A limestone plate with a negative map of Moosburg in Bavaria is prepared for a lithography print.

Limestone was most popular in the late 19th and early 20th centuries. Train stations, banks and other structures from that era are normally made of limestone. It is used as a facade on some skyscrapers, but only in thin plates for covering, rather than solid blocks. In the United States, Indiana, most notably the Bloomington area, has long been a source of high quality quarried limestone, called Indiana limestone. Many famous buildings in London are built from Portland limestone.

Limestone was also a very popular building block in the Middle Ages in the areas where it occurred, since it is hard, durable, and commonly occurs in easily accessible surface exposures. Many medieval churches and castles in Europe are made of limestone. Beer stone was a popular kind of limestone for medieval buildings in southern England.

Limestones and (to a lesser extent) marble are reactive to acid solutions, making acid rain a significant problem to the preservation of artifacts made from this stone. Many limestone statues and building surfaces have suffered severe damage due to acid rain. Acid-based cleaning chemicals can also etch limestone, which should only be cleaned with a neutral or mild alkaline-based cleaner.

Other uses include:

- It is the raw material for the manufacture of quicklime (calcium oxide), slaked lime (calcium hydroxide), cement and mortar.

- Pulverized limestone is used as a soil conditioner to neutralize acidic soils (agricultural lime).

- Is crushed for use as aggregate—the solid base for many roads as well as in asphalt concrete.

- Geological formations of limestone are among the best petroleum reservoirs.

- As a reagent in flue-gas desulfurization, it reacts with sulfur dioxide for air pollution control.

- Glass making, in some circumstances, uses limestone.

- It is added to toothpaste, paper, plastics, paint, tiles, and other materials as both white pigment and cheap filler.

- It can suppress methane explosions in underground coal mines.

- Purified, it is added to bread and cereals as a source of calcium.

- Calcium levels in livestock feed are supplemented with it, such as for poultry (when ground up).

- It can be used for remineralizing and increasing the alkalinity of purified water to prevent pipe corrosion and to restore essential nutrient levels.

- Used in blast furnaces, limestone binds with silica and other impurities to remove them from the iron.

- It is often found in medicines and cosmetics.

- It is used in sculptures because of its suitability for carving.

Travertine

Travertine is a natural, sedimentary stone often used as a building material and found in hot and cold mineral springs around the world. It is most commonly found in hot springs, but can be found in cold springs and limestone caves (where it can form stalagmites and stalactites). It is technically a type of limestone that is formed through a process that includes cycles of water dissolving deposits and evaporation leaving behind layers of aragonite and calcite.

These calcium carbonate layers continue to form over thousands of years to give us the beautiful terraces and cave formations we see today in countries around the world, including Iran, China, Turkey, Hungary, Guatemala, Spain and the U.S.A (at Mammoth Hot Springs in Yellowstone National Park).

While it is a member of the limestone family, it should be noted that travertine and limestone are not interchangeable in the world of architecture, interior design or landscape design, and the names of these types of stones should not be used interchangeably.

Limestone and travertine used together in some applications, particularly in interior design where you might see limestone counter tops paired with travertine floors.

Historically, travertine has been used as a building material for its durability, strength and potential for structural integrity.

For example, when the Colosseum was being constructed, they chose travertine for the external wall, ground floor and main pillars. Another example is the Burghausen Castle in Upper Bavaria, which was constructed almost entirely out of travertine.

Of course, if the Romans were choosing this stone for their bath complexes and the Bavarians were building entire castles out of it, we know that it was not just the strength and longevity that drew them to travertine.

Today, we still use this natural stone for building, but it is much more common — particularly in

Southern California — to see travertine used for floors, patios, pool decks, bathroom and kitchen walls, and other applications where homeowners want both durability and beauty.

It is a very common feature in higher-end homes and is one of a handful of materials (granite, marble, onyx, paving stones, etc.) that prospective buyers ask for by name and that realtors hope are in the homes they are trying to sell.

While granite and marble continue to be popular, travertine is pulling ahead of the pack, particularly because it is lighter than both of these heavy stones, it is abundant, it is fairly easy to quarry and its formation near the earth's surface makes if more weather resistant than stone that is formed deep in the earth.

Travertine Colors and Finishes

Travertine comes in a wide variety of colors, and the names used to describe these colors can vary between suppliers. The most common colors we will see used in residential and commercial applications are ivory, beige, walnut, silver, noche and gold. Travertine's neutral tones can range from a very light ivory to a rich, dark brown, and can have pink, yellow, red, green or gray hues.

Like other natural stones, the color varies based on where the travertine is quarried and which minerals are most predominant in that area.

The finishes we are mostly likely to see when choosing travertine are tumbled, honed, brushed and polished. Tumbled travertine is most common for patios, walkways, pool decks and other outdoor applications. This finish offers a natural, old-world look with gently rounded edges and a textured feel.

These unfilled pavers maintain their porous quality and texture, which makes them a particularly good choice where a durable, slip-resistant surface is needed, such as around pools.

Brushed tiles are similar to tumbled tiles; however, they do not have the rounded edges. The chiseled — or brushed — edges give these tiles an even more rustic look and are usually used for outdoor applications.

Honed travertine is often found in kitchens and bathrooms or for use as interior floors. Honed tiles have a smooth, matte finish and the pores have been filled, which makes them more stain resistant.

Polished travertine tiles are smooth, shiny and beautiful, but they can also be slippery when wet. These are also more stain resistant than brushed or tumbled travertine, making them a popular choice in commercial settings.

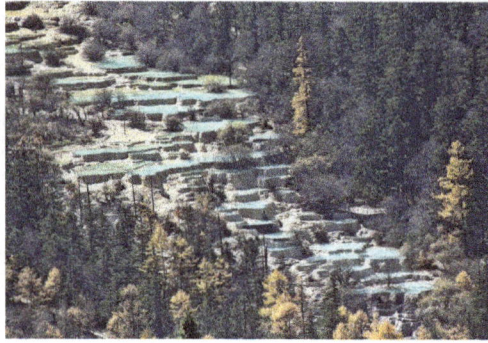

Advantages of Travertine

One of the main reasons travertine is so popular is its versatility, which allows it to be used in a wide variety of applications, including:

- Interior floors

- Interior and exterior walls

- Patios

- Walkways

- Pool decks

- Sitting walls

- Water features

- Counter tops

- Back-splashes

- Shower stalls and bathtub surrounds

- Driveways and parking areas

- Indoor and outdoor fireplace surrounds

It is also durable, resists weathering when used outdoors, low maintenance and beautiful.

Additionally, it is a natural stone option, which appeals to a growing number of today's homeowners.

Its unique look and texture allow travertine to be used in applications where a rustic, old-world look is required, as well as installations where a more modern appearance is the goal.

It is an elegant, attractive addition to indoor and outdoor installations and provides a functional surface for outdoor applications where it is important to have surfaces that are skid resistant, slip resistant and cool to the touch.

This makes them a particularly popular choice for pool decks where travertine provides a long-wearing surface that is slip resistant even when wet and that stays cool even in the afternoon sun.

It is also an ideal choice for homes or businesses with indoor areas that flow into outdoor spaces. For example, if we have a dining room with French doors opening to a patio, installing a travertine floor in both spaces allows we to tie together these living areas for seamless indoor-outdoor entertaining.

Its beauty, durability and versatility are reasons enough to choose travertine for indoor and outdoor applications, but there is also something to be said for the fact that the longest castle in the world (Burghausen Castle, built more than 1,000 years ago) was constructed of travertine, and that the largest building in the world made primarily of travertine (the Colosseum) was constructed in 70 AD.

We do not have to seal travertine tile or pavers, particularly if they are polished, but it is a good idea and is highly recommended. It can be a little confusing for folks, since travertine is known for being so durable and resistant to wearing.

But there is a difference between walking on a natural stone surface and spilling red wine on it, or having your car leak oil on it, or allowing water from your salt water pool to gather on it in puddles.

For the longevity of your investment and your ongoing satisfaction with your installation, it is best to have your installers seal your travertine and to continue to seal it regularly in the future.

This is true for both indoor and outdoor applications using any type of travertine — except polished travertine, which is pretty stain resistant on its own but can still be sealed to make cleanups even easier.

Talk to travertine paving stone installer to determine which type of natural stone sealant is the best choice for your particular situation.

There are impregnating sealers that help retain the natural appearance of your stone, as well as enhancing sealers that bring out the contrasts and colors.

Caring for Travertine

Travertine, marble and limestone often get a bad rap when it comes to their long term care, but these popular stone products are really not that difficult to care for once we know what we are doing.

Mostly, we just need to remember that you are dealing with natural stone and not to use conventional cleaning products, which can etch the surface. Acids and abrasives of any kind can etch travertine, so it is important to also not use natural cleaning products with vinegar, orange, lemon or mild abrasives. It is best to simply wipe down travertine surfaces with a cloth or sponge with warm or hot water, and to dry mop travertine floors, walkways and patios. If something acidic spills on your travertine (even if it is sealed) we will want to clean it up right away. This includes tomato sauce, coffee, wine, sodas and fruit juices. Blot with a soft cloth or sponge until the majority of the liquid is gone, and then wipe the surface with hot water using a cloth or sponge.

When we need or want a deeper cleaning than water can offer, we can use a product designed for cleaning natural stone, which we can find at most hardware stores and home improvement stores.If your travertine gets etched despite best efforts to use coasters and catch spills as they occur, all is not lost.

We may be able to remove the etching with a store-bought, etch-removal product, or we can bring in a professional to repair the damaged area.

Because it is natural stone, it can also absorb oil (or blood or other organics), but even these stains can be removed by using a poultice-like cleaning substance that pulls the oil from the stone.

Oil-removing cleaners are also available at hardware stores and home improvement stores.

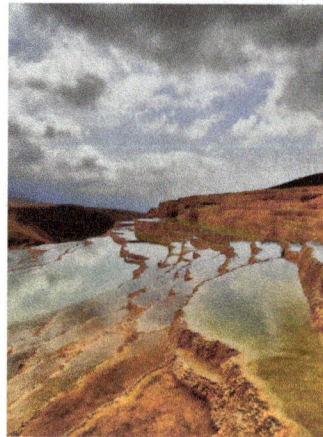

Cast Stone

Cast stone is a highly refined architectural precast building stone manufactured to simulate natural cut stone.

Properly manufactured cast stone is dense and well consolidated. The combination of low absorption and high-compressive strength makes the material generally durable and resistant to freezing-and-thawing distress. The compressive strength of cast stone is usually far greater than is necessary for the application; however, it can serve as an indicator of good quality control and future durability. Cast stone with inferior physical properties, though perhaps adequate for the particular application, may not possess the same service life of a higher quality material whose physical properties are consistent with cast stone industry recommendations.

The Cast Stone Institute is an organization composed primarily of cast stone fabricators and other construction professionals involved in the specification, manufacture, and use of cast stone. According to the Cast Stone Institute's Technical Manual with Case Histories, fourth edition, "The purpose of the Cast Stone Institute is to improve the quality of cast stone and to disseminate information regarding its use."

As a material, cast stone is really a variation of precast concrete. Besides sharing common constituents, cast stone is typically mixed, formed, cured, and stored in a plant environment like precast concrete, which enables rapid, consistent, and controlled fabrication. As with other concrete products, cast stone can be reinforced to increase its ability to withstand flexural and tensile loads. Despite its many similarities, cast stone does differ from precast concrete in a few ways: the mixtures integrate finer aggregates to more closely simulate the appearance of natural stone, the method of fabrication can involve very little water, and the product is virtually always used in nonstructural applications.

Cast stone can also be subject to similar quality control concerns as precast concrete. These can include a lack of consistency in mixture design causing variations in appearance, premature cracking as a result of inadequate curing or insufficient strength gain prior to form stripping, damage as a result of impact during storage, transport or erection, and contaminants or reactive aggregates in the raw materials that can cause internal distress. When properly fabricated, cast stone can be a durable and cost-effective substitute for natural stone, but it may not always look like natural stone. Over time, cast stone can develop characteristics such as cracks, crazing, and discoloration that make it appear less like natural stone as it ages. If quality control is poor, these defects can be more apparent and appear earlier in the service life of the material. Manufacturers should be candid with architects and owners about the potential risks associated with cast stone. In essence, it is a good substitute for natural stone, but not an equal.

Fabrication

Two processes are typically used to produce cast stone: the vibratory dry tamping (VDT) method and wet casting. Both have potential advantages and disadvantages.

The VDT method is unique to cast stone fabrication. To achieve the appearance of natural stone, very dry mixtures of fine aggregates, cement, and water are pounded or compressed into a form on the side that will become exposed in the finished structure. This material is referred to as the face mixture. Depending on the depth of the face mixture required and the complexity or relief of the form, the face mixture is placed in layers called lifts to ensure full compaction of the material into the form. A backup mixture, consisting of coarser aggregate, cement, and water is then poured or rammed into place over the face mixture to fill the remaining portion of the form. The material is allowed to harden and cure, and then the form is stripped and the material stored until it is needed on the project site.

Cast stone produced using the VDT method can replicate stone quite accurately and is less susceptible to surface disruption as a result of free water against the form. Quality control, however, is critical to maintain consistency of both the face and backup mixtures. Changes in thickness of the face mixture can result in variations in density and appearance of the face mixture, as well as cracks due to differential shrinkage between the drier face mixture and the wetter backup mixture.

Backup mixtures are usually highly variable in content because they are not visible when the finished product is used.

An example of cast stone used as ornament at quoins, belt course, and window surrounds on this circa 1920s building.

Wet casting of cast stone is virtually identical to the process used for precast concrete: a form is constructed and then filled with a mixture of aggregates, cement, some additives, and water. Some wet-cast methods can involve multiple lifts of material, or variations between the face mixture and backup mixture. It is allowed to harden and cure for a period of time, and then the form is removed. The formed product is then stored until it is transported to the building site where it will be used. Its principle advantages include greater consistency in physical properties through the material's thickness and better quality control of the material. Its principle disadvantages include lower production rates due to the longer curing time required before stripping, susceptibility to plastic drying shrinkage if not properly cured, and disruption of the finished faces as a result of trapped water at the form/mixture interface.

Curing methods for either technique are also highly variable depending on the cast stone manufacturer. Some cure their product using water misting, steam, curing compounds, or damp curing. The amount of curing also varies, depending on the fabrication process and storage practices, as well as the demand for the product on the job site. If the cast stone is insufficiently cured, then it can experience excessive shrinkage, causing cracking of the surface and increased water absorption.

Common Problems

There are a number of common problems that can occur with cast stone. Whereas some of these occur in cast stone produced using both wet casting and the VDT method, the majority of problems observed in modern construction are associated with the VDT method of manufacture. Unless specifically stated otherwise, these discussions will focus on cast stone created using the VDT method.

These problems range in importance from those that may simply affect appearance or accelerate the need for routine maintenance to those that impact the structural integrity of the material and put the public at risk. Several of the more commonly observed deficiencies found with cast stone are discussed in the following.

Excessive Soiling

Excessive soiling of cast stone surfaces can result from exposure to pollution, soot, and airborne dust. It can also be a result of these materials washing down from other adjacent building surfaces onto the cast stone. Because cast stone is absorptive as well as somewhat rough in texture, particulates can settle into the cast stone surface, or be deposited there by water. Cast stone with higher absorption and lower surface densities can become soiled more quickly since the surface structure is more open.

Crazing

Crazing, or craze cracking, is a network of inter- connected hairline cracks. These cracks usually extend only a few millimeters into the cast stone; however, severe crazing can merge together to form deeper cracks that can allow moisture to reach the interior and, in extreme cases, cause loss of strength or instability. At the very least, these cracks take in moisture and dirt, causing them to discolor. Despite the objectionable appearance and potential for more severe damage, crazing is considered a nonstructural concern and not cause for rejection of cast stone by CSI.

Heavily soiled cast stone at a water table due to water run-off.

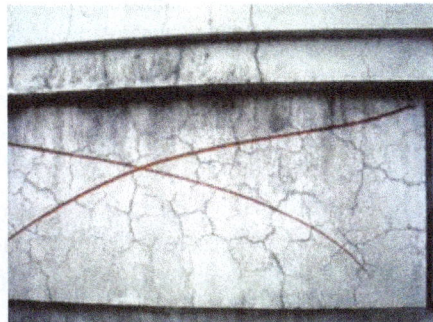

Heavy crazing in a cast stone façade panel.

Crazing is thought to be generally caused by shrinkage occurring at the outermost surface prior to the interior portions of the piece. It can be attributed to curing practices, variable cement content at the surface, excessive wetting and drying, or inadequate ventilation behind the cast stone. The process of tamping also contributes to crazing by creating centers of high compaction (where the tamper impacted the material) surrounded by rings of lower compaction. Crazing appears to be more concentrated in the areas of lower compaction or density. The amount of crazing is also more prevalent at locations where the face mixture is thin. The variations in thickness lead to differential drying shrinkage in the face mixture, as well as variations in density that leads to the formation of craze cracking at the surface.

Racking is probably the most common problem associated with cast stone. CSI recommends that pieces containing cracks in excess of 0.005 in. (0.127 mm) not be accepted in a quality cast stone installation. Cracking can develop as a result of many different conditions and range in impact from cosmetic to a loss of structural capacity. Several of these are described in the following.

Restraint of Volume Change

Often cast stone is rigidly attached to the backup structure for support, with no allowance for volume change of the material. Although VDT-cast stone is manufactured with low water content and experiences less shrinkage than wet-cast products, shrinkage does occur and continues for several years after fabrication. The cast stone is also subject to volume changes due to thermal cycling and will typically experience a greater temperature swing than the unexposed backup structure whose temperature range in service is often moderated by the thermally controlled building interior. If the ends of the cast stone are restrained, the differential volume change between the cast stone and the backup structure results in cracks forming across the face of the cast stone.

Crack resulting from restraint from shrinkage.

Some manufacturers will attempt to control cracking resulting from volume changes by introducing reinforcing steel in both directions across the face of the cast stone piece. Unfortunately, this is often ineffective because of the difficulty in achieving adequate consolidation of the material around the reinforcing steel to control and distribute the cracking. Poor consolidation around reinforcing steel in the transverse direction (perpendicular to its span) can actually form weak planes through the thickness where cracks are more likely to form.

Cracking that resulted from corrosion of reinforcing steel.

Corrosion

Corrosion of embedded reinforcing steel can lead to cracking at the surface of the cast stone. This type of cracking is often accompanied by delamination of the material at the depth of the reinforcement, leading to further instability of the cast stone in the form of spalls.

Insufficient Strength

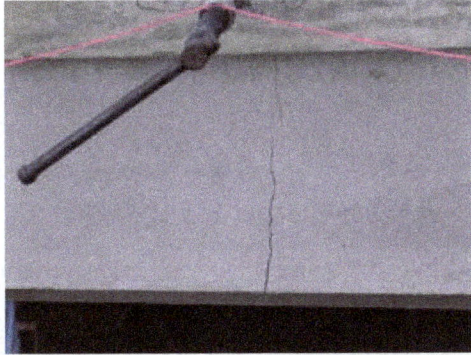

This crack was formed as a result of excessive loading in flexure.

Although cast stone is typically not heavily loaded in most building applications, pieces can develop relatively high flexural stresses if they are spanning across openings, not fully bedded along their length, or in any other orientation with just two points of support. The flexural stresses are exacerbated if the piece is long and slender. Flexural cracking will typically form in the middle third of the span and run from the top to bottom edge of a horizontally oriented piece or across a piece that is oriented vertically, such as a window jamb.

Poor storage and handling cast stone on site can result in improper loading and damage.

Handling

Cracks can also develop as a result of mishandling or unintended loading during transportation or erection. Cast stone is usually stripped shortly after forming, moved to a curing facility, moved for cleaning, and moved again to yard storage. It will often be transported to the project site prior to gaining full strength, making it more susceptible to damage while being loaded and unloaded. Cracks can develop if a piece is picked up or stored in a manner not intended many cast stone fabricators will furnish lifting hardware cast in larger pieces to avoid damage during lifting operations.

Patching Failures

Patching is the process of repairing a spall or chip in cast stone by placing into the defect a fresh, formable cementitious material mixed to match the damaged cast stone as closely as possible. Patching is usually performed by the manufacturer of the cast stone, after the piece is installed.

A patch relying solely on cementitious bond in an overhead condition.
Note the failure of the feathered edge at left.

Despite duplicating the cast stone components and their proportions, patches rarely match in all environmental conditions because the density and absorption of the original material cannot be replicated when the patch is installed. Because it is so difficult to obtain a good match between the patching material and the cast stone, only damage truly noticeable should be repaired. Pieces with spalls or chips greater than 8 in. (203 mm) square should be replaced. No matter the size, an acceptable patch must not be visible from more than 20 ft (6 m) away.

Patches that have failed and have become unstable should be removed.

Patches fail by shrinking excessively or debonding from the substrate. Water can work between the patch and the cast stone substrate through separations at the bondline resulting from shrinkage; water can further degrade the patching material, become trapped and freeze, promote efflorescence, or simply make the patch more visible and detract from the overall appearance of the cast stone. Unstable patches that are debonded from the substrate or severely cracked should be removed to prevent them from falling out on their own, particularly where pedestrians or traffic could be impacted.

Corrosion of Reinforcing Steel

Whereas most cast stone is not used in load- bearing applications and does not require reinforcement, reinforcing steel or reinforcing bar can be used to increase its strength in flexure or enhance its ability to be handled or transported without damage. Placement of reinforcing steel is of particular concern in VDT-cast stone panels and should be avoided. Usually the face mixture is placed first, then the steel is set, and the backup mixture poured around it. Therefore, this process does not allow for the reinforcing steel to engage the face mixture. If the backup mixture is dry - tamped as well, it is nearly impossible to achieve adequate consolidation around the reinforcing steel that is sufficient to develop its strength.

If it must be used, wet-cast methods of fabrication are preferable so that the reinforcement can be fully encapsulated in cementitious material. It is also important to provide adequate cover over the steel to increase the time it takes for carbonation to reach the depth of the steel. Reinforcement that is less susceptible to corrosion, such as galvanized or epoxy-coated bars, also help to reduce the risk of corrosion-induced distress.

Spalling

Spalling and incipient spalling can occur for a multitude of reasons. Spalls can develop at anchor points where stresses are high and the cast stone is cut to receive the anchors. It can also occur as a result of setting procedures—pry bars are often used to position the stone, and the weaker edges and corners can break due to the applied pressure.

Delamination/Separation of Lifts

Cast stone manufactured using the VDT method is compacted into forms as layers called lifts. These lifts are intended to bond to each other, with the tamper forcing the layers into intimate contact. When the material is spread in the forms, however, the material being the most highly compacted is that closest to the tamper and furthest away from the layer below. This creates zones of lower compaction at the lift lines that can be more absorptive and break down more quickly if exposed to the environment; exposed lift lines can take in a substantial amount of water and erode, leaving fissures on the surface that are visually unappealing and increase the amount of water able to reach the interior of the cast stone.

Often bond at the interface between lifts is lacking. Bond can also be reduced over time if the lift inter- face is exposed to the environment. Without adequate bond or mechanical engagement, the outer lift (often the face mixture) can separate from the back- up mixture and become unstable. This condition, in combination with the presence of cracks in the face mixture, can allow portions of the face mixture to fragment and spall.

Solutions

Over the years, many creative approaches have been developed to restore, repair, and maintain cast stone. The cast stone industry and professionals engaged to correct deficiencies in building materials have developed repair materials and methods to address many of the problems previously mentioned. Some of the more common repair/maintenance approaches are discussed in the following.

Delamination of the face mixture from a large spandrel panel.
Round patches are patches from prior core sampling.

Cleaning

Most soiling can be treated successfully with conventional water rinses, detergents, or chemical cleaners. The use of more aggressive cleaners, such as those containing acids, should be avoided or used judiciously since they can dissolve the cementitious binders in the material and lead to erosion and roughness. Cleaning with high-pressure water (greater than 300 psi [2 MPa]) should also be avoided as it can remove the paste surrounding the aggregate, roughening the surface. If the surface of the cast stone is rough or cracks are present, more debris is retained, making the material "dirtier" in appearance. Cracks are more difficult to clean because the soiling material is drawn more deeply into the crack where conventional surface cleaning may not reach.

Water Repellent Application

The application of penetrating water repellents such as silane and siloxane blends to cast stone can reduce its absorption and improve its resistance to soiling by making the surface hydrophobic and less able to absorb contaminants deposited by water. Low water absorption is critical to maintain durability, reduce the appearance of crazing, and reduce soiling overall.

If cast stone is exhibiting visible crazing, water repellents can be applied after cleaning to help prevent the crazing from becoming more pronounced. It prevents contaminants from being redeposited in the cracks and also prevents water from wicking into the body of the cast stone.

Re-etching

When originally fabricated, a cleaning solution most often containing muriatic acid is used to remove the excess paste at the surface and to expose the brighter stone aggregate. Occasionally, if soiling is severe, or if crazing is visible and darkened by contaminants filling the surface of the fine cracks, a similar acid-based wash can be used to improve the appearance. The stronger cleaning solution aggressively attacks the material in the cracks and surface irregularities. The author's experience suggests that the appearance of shallow crazing can be improved with this method and is worth attempting; however, older crazing that penetrates more deeply into the surface is not typically improved by the application.

One must also consider, however, that it is far more difficult to apply an acid wash to cast stone once it is in place, particularly if it is oriented in a vertical position. Adjacent surfaces often must be protected from damage by the caustic cleaners, and run-off must be collected and neutralized or otherwise controlled.

Architectural Coating

When the cast stone is severely crazed, soiled, discolored, or contains a number of poorly matched patches, an architectural coating or pigmented sealer can be an attractive option. Although the original appearance simulating natural stone is lost, coating and pigmented sealers offer a consistent, fresh appearance Coatings will bridge small cracks and surface irregularities, and provide a water-resistant finish for the cast stone, reducing future concerns about absorption.

Coatings can range in formulation from acrylic elastomeric to potassium silicate-based materials. The most critical characteristic is breathability, or its ability to allow water vapor to pass from the cast stone to the exterior. Coatings that are not sufficiently breathable will trap moisture, peel, blister, and encourage freezing-and-thawing deterioration within the cast stone. Coatings and sealers do require reapplication; our experience suggests recoating should be anticipated every 5 to 10 years, depending on the product and its environmental exposure. Most coatings can be easily cleaned with mild detergents.

Crack Treatments

Cracks that are nonstructural but can allow excessive water penetration to reach the interior of the cast stone should be sealed. Cast stone producers will often rub a cementitious slurry or grout into the crack, filling up the surface; however, the crack quickly reforms through the thin brittle application. To more successfully seal cracks, the surface of the crack should be widened and deepened to accept an appropriate amount of material. Elastomeric sealant, cementitious grout, and structural adhesives have all been used with mixed results. The more rigid materials have a more pleasing appearance but can fail even if the crack is considered stable. Sealants are more forgiving to movement but can be more visible due to the textural differences between them and the surrounding cast stone.

Patching

Example of architectural coating applied over supplemental anchors

Patching of cast stone is inevitable. Often, the first priority for the manufacturer is to install a patch that minimizes the impact to the cast stone and matches well. Unfortunately, these patches often fail due to poor surface preparation and reliance on bond strength of the patch material to hold it in place. Because the patch will shrink after its placement in the cast stone, the bond can be broken as a result of the volume change. The manufacturer will also typically taper the patching material out to the edge of the chip or spall, producing a thin, fragile edge. These feathered edges do not have the integrity to stay bonded and break off, leaving the rest of the patch vulnerable to increased water penetration.

Proper patch installation must make compromises in the appearance. A spall must sometimes be broadened and deepened so that the patch material will be firmly engaged into the surface. The edges must be cut to eliminate feather edging, and mechanical anchoring is necessary to ensure that if the patch does lose bond, it will stay engaged in the substrate. These practices will produce a patch that is more noticeable, but one that will be far more durable.

Preparation for patch repairs. Note the saw cut edges and supplemental
anchors to engage the new patch material

Although many improvements have been made recently regarding the quality of cast stone as a result of more stringent requirements for quality control in the industry, there are still some areas where the standard specification and industry requirements could be improved to better ensure a quality product will be delivered and the material's end user will be satisfied.

Cast stone is a unique material that offers modern designers the appearance of natural stone, but with all the advantages of a manufactured product. Conversely, proper manufacturing processes and quality control are critical to providing a good cast stone product. A better understanding of the material's advantages and limitations are essential to make certain that all parties involved in the cast stone application are pleased with the final installation.

Cobblestone

Dotting the streets of historic towns and beloved in meandering garden paths, cobblestones are some of the most recognizable (and charming!) stones used in hardscaping. Simply put, cobbles are rounded stones that are traditionally used to pave roads and paths. They are usually between two and ten inches in diameter or length and are often taken from rivers, where the constant running water gradually wears away at the stone and forms those signature rounded edges.

The word cobblestone is derived from the English word "cob", which means something round or lumpy. For thousands of years, they've been used in many different societies to create paved roads, which were a huge step up from dirt paths. Doesn't sound like much an improvement, you might think? Sure, by today's standards cobblestones can be lumpy and difficult to walk on. However, dirt roads were very easily washed out and rendered impassable by rain or floods, as the thick mud that was produced would easily stop horses and carts.

Cobblestones helped make early roads much easier to pass and helped pave the way for easier transportation of people and goods. The stones were usually set into a layer of sand and bound with mortar to create a relatively even surface that easily allowed carts, carriages, and horses to pass quickly and easily, all while avoiding the creation of mud and dust in inclement weather. In addition, the hard cobbles could withstand heavy traffic for decades and even centuries without needing replacement or repair — in fact, many historic cities still retain at least some of their original cobblestone roads and paths.

In addition to roads, cobblestones have historically been used to decorate buildings, walls, and fireplace hearths. And, while rounded, oval-shaped cobbles are what most immediately come to mind, they can also be square and rectangular. However, they always retain their rounded, irregular edges. While today homes and hardscaping features designed with these materials create a quaint, old-world feel, in earlier times they were simply the look of the day and primarily a way to make homes, walls, and roads a bit sturdier.

Cobblestones are still in use today primarily as a decorative accent. Although they were a sturdy improvement over dirt roads back in the day and they don't warp or crack like modern asphalt, don't expect to see a return to cobblestone roads today — they're difficult and expensive to maintain and the bumpy, uneven texture causes a lot of wear and tear on modern cars. Still, cobbles are a popular choice for homeowners looking to capitalize on that old-world charm and are usually used to create pathways in the front and backyards, and they can be used to create beautiful retaining walls for gardening. They're also still a very popular choice for fireplace hearths, as the charming, natural imperfections can give a room a little rustic elegance.

When shopping for cobblestones, there are tons of options. Antique cobblestones, reclaimed from real old-world buildings and paths, can be purchased and reused in your own hardscaping projects though they are often quite expensive. If you're buying new cobbles, you can get them in nearly any stone you'd like, such as granite, limestone, and basalt. They also come in a wide variety of colors, such as red, green, gray, black, pink, tan, brown, and variations thereof. And, as we said above, think outside the oval-shaped stones! Square and rectangular cobbles can create beautiful outdoor figures, and many homeowners like the more regular patterns they can create. However, a popular recent trend is to use cobblestones in different colors to create patterns, such as swirls, zig-zags, and other simple designs.

There's a reason cobblestones have been used for ages — they're sturdy and have a simple, everyday beauty.

References

- Pepper, Terry (2003-03-25). "The story of Cream City brick". Seeing the Light: Lighthouses of the western Great Lakes. Retrieved 2006-10-04
- Bricks-types-uses-and-advantages-844819: thebalancesmb.com, Retrieved 16 May 2018
- Mudbrick, building-materials, interactive-green-building-guide: builditbackgreen.org, Retrieved 30 June 2018

- London-yellow-stock-brick: lrbm.com, Retrieved 19 July 2018

- Marble, rocks: geology.com, Retrieved 20 April 2018

- Paulsen, Eric (2004-11-19). "Cream City brick built Milwaukee's name". OnMilwaukee.com Travel and Visitor's Guide. Retrieved 2006-10-04

- What-is-travertine: installitdirect.com, Retrieved 18 March 2018

- What-are-cobblestones, building: lgsgranite.com, Retrieved 21 May 2018

Moisture Protection in Construction

Structural dampness is one of the most common problems in construction. It is the intrusion of undesirable moisture in the structure and foundation of a building. This chapter has been carefully written to provide an understanding of the techniques of ensuring moisture protection in construction, with a close examination of the fundamental concepts of structural dampness, damp proofing, housewrap, basement waterproofing, etc.

Building Envelope

The building envelope is the physical barrier between the exterior and interior environments enclosing a structure. Generally, the building envelope is comprised of a series of components and systems that protect the interior space from the effects of the environment like precipitation, wind, temperature, humidity, and ultraviolet radiation. The internal environment is comprised of the occupants, furnishings, building materials, lighting, machinery, equipment, and the HVAC (heating, ventilation and air conditioning) system.

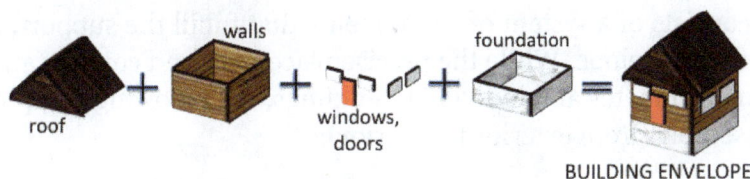

Figure. The components of the building envelope

Improving the building envelope of houses is one of the best ways to get better energy efficiency.

Function

A building envelope serves many functions. These functions can be divided into 3 categories:

- Support: to ensure strength and rigidity; providing structural support against internal and external loads and forces.
- Control: to control the exchange of water, air, condensation and heat between the interior and exterior of the building.
- Finish: this is for aesthetic purposes. To make the building look attractive while still performing support and control functions.

Physical Components

The building envelope includes the materials that comprise the foundation, wall assembly, roofing systems, glazing, doors, and any other penetrations. The connections and compatibility between these elements is critical to ensure that the building envelope functions as intended.

Foundation

The foundation is the structural component that transmits the loads from the building to the underlying substrate. Typically, some combination of reinforced concrete walls, slabs, and footings constitute the structural components of the foundation. However, the foundation must also be designed to control the transfer of moisture and thermal energy into the interior space.

The transfer of thermal energy through the foundation can be controlled by providing insulation between the interior and exterior environments; however, in some cases the foundation insulation is neglected to reduce construction costs.

Waterproofing the foundation is typically completed by applying a liquid applied asphaltic dampproofing. Additional waterproofing products such as sheet-applied membranes, liquid membranes, cementitious waterproofing, and built-up systems are also viable options.

Drainage around the perimeter of the foundation must be provided to prevent long-term underwater submersion of the waterproofing membrane. One example of a perimeter foundation drain is weeping tile placed in trench complete with gravel ballast backfill, also known as a french drain. In some cases, a sump pit and pump system will be required in addition to the perimeter drain.

Wall Assembly

The wall assembly consists of a system of components that fulfill the support, control, and finish function of the building envelope. While the precise placement and configuration of each component may vary between climates and individual buildings, the following components are typically found in the wall assembly (from exterior to interior):

- Exterior cladding
- Exterior sheathing membrane
- Exterior sheathing
- Insulation
- Structural components
- Vapour barrier
- Interior sheathing

Roofing System

The roofing system is an important part of any house, as it keeps weather out. It consists of shingles on the outside, which are on top of tar sheeting as a vapor barrier. Inside of the tar paper is wood sheathing. Beyond this, the attic areas in most houses are insulated with fiberglass spray insulation. It tends to be fluffy, pink fiberglass. Inhaling fiberglass is extremely bad for a person's respiratory system, so it is important to wear a mask if this insulation type is in one's roofing system.

Glazing

Glazing refers to the panels in windows, doors and skylights - usually glass - that let light through.

Door

Doors are included in the housing envelope, as they tend to be the biggest holes in the envelope. Having outer doors that seal well drastically improves the thermal efficiency of a house.

Other Penetrations

These may include a chimney, or vents for a dryer or stove.

Damp

Structural dampness is the presence of unwanted moisture in the structure of a building, either the result penetration from the outside or condensation within structures. A high proportion of damp problems in buildings are caused by condensation, rain penetration or rising damp.

Breathability and Walls

Old buildings must be allowed to breathe. Whereas modern buildings rely on keeping water out with a system of barriers, buildings that pre-date the mid-19th century are usually constructed of absorbent materials that allow any moisture that enters to evaporate back out.

Because most old buildings were constructed with solid walls without damp proof courses and originally had no roofing felt, rain or below ground moisture could both enter. This did not, however, mean dampness was inevitable. Before central heating was commonplace, heat from open fires drew in large quantities of air through loosely fitting windows and doors. This high rate of ventilation would have quickly evaporated moisture from permeable internal surfaces while the wind dried out any damp roof timbers or permeable external wall surfaces.

An equilibrium was therefore established whereby the moisture being absorbed was equal to that evaporating. When upgrading an old building, you must maintain this equilibrium for the building to work as intended and remain dry.

Causes of Damp

The main risks arise from:

Air Moisture Condensation

Energy-saving measures that reduce ventilation in old buildings such as double-glazing increase relative humidity. Humidity is also raised by modern lifestyles that generate large quantities of water vapour, from bathing, cooking and washing. Condensation will occur on any surface below the dew point (i.e. temperature at which saturated air releases surplus moisture vapour). Interstitial condensation within the pores of materials reduces thermal insulation and further increases the risk of condensation.

Penetrating Damp

Roofs, chimneys, parapets and other exposed parts of a building are most susceptible to rain penetration, especially where access for maintenance is difficult. Junctions in roofs are potential trouble spots, with water exploiting defective lead flashings, mortar fillets, ridges or hips.

Concentrated and prolonged wetting of walls and external joinery arises from poorly maintained rainwater fittings, and leaks from parapet and valley gutters can cause significant damage to structural roof timbers. Hairline cracks in pointing and render invariably admit moisture where cement mortar has been used for repair, rather than lime.

Internal Spillage

This results from overflowing baths or showers, burst pipes, the gradual breakdown of pipe joints, leaks from washing machines or dishwashers, and accidental damage.

Below Ground Damp

This may be rising damp, which is neither as widespread as commonly thought nor a total myth, as sometimes now claimed. Floors become damp where the evaporation of moisture from below is inhibited by vinyl sheet, rubber-backed carpets or other impervious coverings.

New concrete floors or impervious coverings also drive excess moisture into the bases of nearby walls (including chimney stacks), where it rises by capillary action. Damp proof courses were not compulsory in walls prior to 1875 but this is only likely to become a problem where breathability is compromised. In addition to rising damp, below ground moisture can result in problems where ground levels around your building rise unduly.

Tips for Diagnosing Damp

Roofs and Rainwater Fittings

Inspect your roof during wet and windy weather to decide if a damp ceiling patch is due to roof leakage and/or condensation. Debris on the ground (broken slates, tiles and so on) or daylight seen inside lofts indicate possible roof problems.

Defective rainwater fittings may be most obvious during heavy rain, but stains on walls and plant growth provide further clues. Don't forget to check gulleys at ground level.

Condensation is diagnosed from diffuse areas of damp, beads of water droplets on hard shiny surfaces and mould growth on internal finishes. It is intermittent, like penetrating damp, but unrelated to wet weather.

Penetrating damp typically shows up as well-defined patches after heavy rain on south- and west-facing walls. Anticipate moisture ingress through hairline cracks in unsuitable hard, modern cement pointing or rendering.

Below ground moisture causing rising damp can extend up to 900 mm above floor level, sometimes with a classic tidemark on finishes. Salts appear as white deposits but mould growth is rare.

Plumbing

Unusually high water bills (if metered) or a constantly refilling tank may suggest leakage.

Damp Proof Courses and Alternative Systems Compared

Retrospective Damp Proof Courses:

Physical:

- Inserted by cutting in or during rebuilding.

- Can cure rising damp but this drastic method is usually inappropriate.

- Drawbacks: possible major structural problems; potential damage to historic finishes internally; unsuitable for randomly coursed walls; access difficulties; deterioration sometimes of masonry below damp proof course where moisture is concentrated.

Chemical:

- Walls impregnated with chemical solution through holes at bottom to create waterproof barrier.

- Widely used today but not always appropriate for old buildings.

- Drawbacks: drilling holes inadvisable in flint, granite, etc.; hard to form proper barrier in rubble walls with voids; holes unsightly; deterioration sometimes of masonry below damp proof course where moisture concentrated.

- Cost: typically £195/m (including replastering).

Ceramic tubes:

- Holes drilled to receive porous siphons approximately 50mm in diameter that absorb damp and evaporate it from each tube.

- Sound in theory but problems may occur in practice.

- Drawbacks: salt accumulation in tubes may increase moisture; air-flow sometimes inadequate; tubes commonly set in hard cement mortar; unsightly.

- Cost: typically 125/m.

Electro-osmosis:

- Electrical potential aimed at reducing capillary rise using electrodes bedded in wall.

- Cheap but dearth of evidence that electro-osmosis is effective and system rarely used today.

- Drawbacks: adjustment of current needed to match variations in damp along a wall usually impractical.

Other:

- An Austrian product presently under trial in the UK claims to inhibit the passage of water up a wall by inducing a local magnetic field. Achieved non-invasively with unit plugged into mains, typically in loft.

- Likely cost: £3,000/unit (one unit covers an average-sized house).

Investigating Damp

Scientific analysis can be an essential aid for accurately diagnosing a damp problem but the importance of your sight, feel and smell should not be undervalued. Tests include the use of electrical resistance and capacitance meters, on-site moisture testers, hygrometers and salt analysis.

Bear in mind though, that care must be taken when interpreting results. A frequent mistake is to diagnose rising damp purely on the basis of high electrical moisture meter readings. Elevated readings may indicate the presence of salts from past dampness or, if there are no salts, an on-going condensation or possible penetrating damp problem. Continued monitoring is often needed to establish the true cause of a damp problem.

Surveyors have a legal duty to follow a trail of suspicion. Regrettably, many still simply note the occurrence of high meter readings and pass on all responsibility for further investigation to remedial treatment contractors. These contractors have a vested commercial interest, encouraging over-specification. Should a mortgage company insist on work you believe is misguided, challenge this and consider obtaining a written report from an independent surveyor or architect.

Remedial Measures

Effective remedial measures depend on accurate diagnosis, but applying staged remedies can also be part of understanding the cause of a damp problem. Before deciding on more extensive work, the first remedy may involve nothing more than basic maintenance such as clearing a blocked rainwater gulley. Remedies will either cure dampness by addressing the cause (for example, improving drainage) or will manage it by treating the symptoms (changing washing or cooking habits, for instance).

Be sceptical of written guarantees, which are often loaded with get-out clauses and may have no insurance backing. The right approach from your contractor coupled with good workmanship is your best guarantee.

Controlling Air Moisture Condensation

Condensation can be treated by reducing air humidity or keeping surfaces above dew point temperature. Humidity is reduced by cutting the amount of moisture available or increasing ventilation by opening windows, etc. Tumble dryers should be vented to the outside if not of the condenser type, and clothes drying indoors is best avoided.

Temperatures are maintained above dew point with suitable heating. The permanent use of dehumidifiers is a poor substitute for efficient heating and adequate ventilation. Condensation in chimney flues can be eliminated with proper linings. Redundant flues that have been sealed should be fitted with ventilation grilles or re-opened. Lofts should be well insulated and ventilated but make sure insulation does not restrict ventilation at the eaves.

Controlling Penetrating Damp

Repair and Maintain the Roof

Reinstate dislodged and missing slates and tiles before damage occurs to roof timbers or plaster ceilings. SPAB, the Society for the Protection of Ancient Buildings, recommend that renovators avoid spray-on roof foams for the underside of roofs, or external bitumen coatings although other experts disagree. SPABs view is that they prevent proper inspection, hinder the re-use of slates or tiles and, by reducing ventilation, increase the risk of decay.

Brush moss off roofs since it can block gutters and retain moisture, which may damage certain roof coverings in frosty weather. Also, clear gutters and rainwater pipes regularly, particularly if your building is surrounded by trees or perched on by pigeons. Parapet and valley gutters need to be cleared of snow to prevent melt-water rising above them and causing damp internally.

Repair and Protect Walls

Re-point deeply eroded mortar joints in walls. Whilst cement is fine for modern buildings, it is important to use a lime: sand mix (preferably without cement) for most buildings pre-dating about 1900. Localised re-pointing is generally all that is required. Daub, lime mortar or oakum (ships caulking) are useful for closing gaps that may develop around the edges of panel infillings in timber-framed buildings.

Where rain penetrates an exposed south- or west-facing wall, lime wash, lime render and slate or tile hanging are traditional solutions although these cannot be employed without changing the external appearance of the wall. Sometimes installation of a ventilated dry lining system internally is appropriate. The use of colourless water-repellent treatments or plastic-based paints on old masonry is strongly inadvisable.

Controlling Below Ground Damp

The best solution to rising damp may well be to take measures that help your building breathe. Replacing hard cement render or pointing using a more suitable lime-based mortar often improves a damp wall and enables rising damp to dry out. Conversely, the application of waterproof renders and bituminous coatings tends to create or exacerbate damp problems.

Where a floor has a modern damp-proof membrane (horizontal barrier or DPM) that is displacing moisture to the bottoms of walls, it may be sensible to replace this completely with a breathable construction or to at least provide a breathing zone for evaporation around the perimeter of the room. When underfloor heating is being installed, there are many situations where DPM-insertion can be avoided by employing materials such as lime concrete and expanded clay insulation.

Reducing or removing the source of moisture may also help alleviate rising damp. French drains can be an effective and relatively inexpensive answer but it is preferable not to site them directly against walls and rodding points must be provided. Otherwise, blockages can effectively convert them into a sump and increase dampness. Consider also the structural and archaeological implications.

Although retrofit DPCs can sometimes be appropriate, with an old building always consider first whether rising damp is actually too minor to matter and, if it is significant, whether more sympathetic ways exist of dealing with it. Where any timber is at risk of decay, for example, you might be able to simply isolate it. Similarly, the eradication of any contributing moisture from other sources such as rainsplash off closely abutting patios could obviate the need for more extensive remedial treatment.

Damp can be particularly troublesome in cellars but increased ventilation (including opening up redundant flues), re-pointing and lowering the water table locally can be effective. Failing this, it may be worth considering a dry lining system. Tanking (applying waterproof linings to walls and floors) is not recommended in old buildings.

Internal Finishes

To minimise the risk of future problems, lime plaster should usually be used for any replastering rather than the anti-sulphate or renovating plasters favoured by many treatment companies. Decoration with paints such as limewash and soft distemper, where possible, will maximise breathability. Old items of joinery removed during work should be carefully repaired and reinstated, not automatically replaced.

Damp Proofing

Damp proofing is a general term that covers methods and treatments used to prevent damp from being absorbed through walls or floors into the interior of a property.

Any property can be subject to damp problems, especially older properties which may have been constructed without a damp-proof membrane. Whether it's rising damp or penetrating damp, the property care specialists are experts at identifying the types of damp within a property as well as potential problems.

Types of Damp Protection

The two types of protection methods for damp proofing residential and commercial properties are Damp Proof Course (DPC) and Damp Proof Membrane (DPM)

Most properties we live in today should have evidence of a damp proof course about 150mm (6inch) above ground level. This may be seen as a slightly thicker mortar course with a slate or bitumen sheet poking through. These damp course materials are fairly resilient, unless movement of the building subsidence causes them to crack.

There is sometimes confusion between what a damp proof membrane is and what a damp proof course is, particularly as they can be used together. We advise linking the two to allow a building to be fully protected from the damp, wet ground around it.

Penetrating Damp

Penetrating damp treatments for your property.

Treat Rising Damp

Treatments for and prevention against rising damp.

Waterproofing & Tanking Solutions

Treatments for basements, cellars and vaults.

Damp Proof Course

A damp proof course (DPC) is one of several damp proofing treatments used to prevent damp problems developing within a property. Damp proof course repair can be applied using a variety of different methods and is a long term solution to preclude moisture from entering a property through walls. The build-up of excess moisture within a building can eventually result in structural damage and therefore pose a risk to your property.

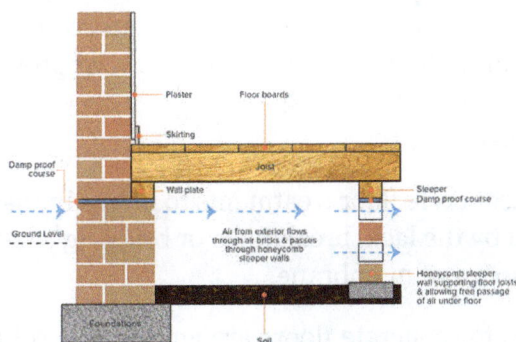

The property care specialists are experts at identifying the types of damp within a property as well as potential problems and will apply the most suitable treatment based on damp proof course regulations. All our injected damp proof course treatments come with a 30 year damp proof course guarantee to ensure a safe and stable property.

Damp Proof Membrane

A damp proof membrane (DPM) is another common method used to prevent rising damp from occurring within a property. The damp proof membrane sheets are made from materials such as polyethylene or butyl rubber and act as a successful barrier to prevent damp from making it's way into a property. The well-qualified technicians have a great understanding in how to resolve damp issues and are experienced in installing damp proof membranes for walls and floors within a property.

Working of Damp Proof Membrane

A sheet of material (impervious to water) laid in one piece that is placed beneath the concrete floor of a property to prevent groundwater seeping upwards through the concrete base. A common example is polyethylene sheeting laid under a concrete slab.

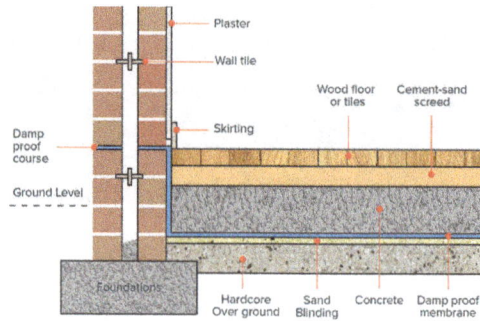

Damp Proofing Walls

The build-up of excess moisture within a building can lead to damp walls which can eventually result into a handful of problems. If you notice signs of damp on walls within your property then we strongly advise you act fast to resolve the issue. Damp on internal walls is unsightly, unhealthy and is the primary cause of wood decay in a building, which can put your property at risk of structural damage.

The property care specialists have many years' experience in treating damp walls and are experts at applying the appropriate damp proof course based on your requirements. Damp proofing internal walls come as a second nature to us so you can be sure your property is in safe hands.

Damp Proofing Concrete Floor

We offer professional damp concrete floor treatments to help eliminate rising damp from within a property. This can be caused by the lack, breakdown or bridging of a physical damp course, or the absence or damage of a damp proof membrane.

The damp proof membranes for concrete floors are an effective solution to counter rising damp and at Rentokil, the property care specialists are registered TrustMark contractors and proud members of the PCA so you know that your property is in safe hands.

Remedial Damp Protection

Issues with your current damp protection can occur due to subsidence or the result of long term deterioration which allows damp to rise through the walls. If you begin to notice:

- Peeling paint or wallpaper
- Discolouration
- Decaying skirting boards and timber floors
- Salt staining
- Crumbling plaster

The Certificated Surveyor in Remedial Treatment can conduct a professional damp survey to identify the cause of damp. Rentokil's remedial damp proof course can then be installed to ensure your property is damp protected.

Interstitial Condensation

Interstitial condensation is very simply, condensation within the structure of a building, i.e. not on the surface and not clearly visible until often, it's too late and the damage is done. Warm air can hold a lot of moisture, so air within a dwelling for example, is perfect for moisture-laden air, especially as we create so much of it – through washing, cooking, breathing, etc. Condensation occurs where this warm, moisture-laden air is suddenly cooled and can no longer hold the high levels of moisture. In a building, this is usually when the warm air meets a cold surface such as a mirror, window or external wall. If the wall is uninsulated, then it can be extremely cold in winter time and condensation will be visible on the surface.

But what happens if the wall is fitted with internal wall insulation (dry lining), for example? The moisture contained in air is water vapour and is usually completely invisible to the human eye, as it is in such tiny particles. This water vapour can easily pass through most building materials such as plaster, slabs, insulation, etc. And so, they do just that.

So, Our wall is insulated on the inside. This makes the external wall even colder, as now the heat cannot get out to heat it up. But the moisture can. Now, the moisture-laden air is cooled even further than before and dumps even more condensation, down along the wall surface, exactly as before, only now, the old wall surface is within the overall structure of the wall (including the dry-lining).

This is called "interstitial" condensation, as it happens within the wall's structure. Any moisture within a structure is not good. It leads, first of all, to nasty moulds and fungi to grow within the structure which leads to very poor air quality within the building. This can have severe health implications. Dampness also leads to structural damage, whether it's rotting timber, degraded insulation, cables or pipes/fitting corroding, etc. It's a huge problem because it can't be seen for maybe several years but by then, it's far too late to do anything about it. This can equally be a problem in new and existing buildings, so great care should be taken, especially when altering the make-up of any building element e.g. wall.

The good news is that correct construction methods such as: using an appropriate "vapour check", a barrier against water vapour; placing the insulation in a different position, such as outside the wall, or introducing the appropriate ventilation where condensation may occur; can eliminate the possibility of interstitial condensation. It is critically important to consider what construction methods are employed before work is carried out. Unfortunately, unscrupulous or ignorant contractors may try to lead we down the wrong path as the incorrect and easy methods are generally much cheaper, quicker and easier to install.

Vapor Barrier

The function of a vapor barrier is to retard the migration of water vapor. Where it is located in an assembly and its permeability is a function of climate, the characteristics of the materials that comprise the assembly and the interior conditions. Vapor barriers are not typically intended to retard the migration of air. That is the function of air barriers.

Confusion on the issue of vapor barriers and air barriers is common. The confusion arises because air often holds a great deal of moisture in the vapor form. When this air moves from location to location due to an air pressure difference, the vapor moves with it. This is a type of migration of water vapor. In the strictest sense air barriers are also vapor barriers when they control the transport of moisture-laden air.

Vapor barriers are also a cold climate artifact that have diffused into other climates more from ignorance than need. The history of cold climate vapor barriers itself is a story based more on personalities than physics. Rose regales readers of this history. It is frightening indeed that construction practices can be so dramatically influenced by so little research and reassuring indeed that the inherent robustness of most building assemblies has been able to tolerate such foolishness.

Problems

Incorrect use of vapor barriers is leading to an increase in moisture related problems. Vapor barriers were originally intended to prevent assemblies from getting wet. However, they often prevent assemblies from drying. Vapor barriers installed on the interior of assemblies prevent assemblies from drying inward. This can be a problem in any air-conditioned enclosure. This can be a problem in any below grade space. This can be a problem when there is also a vapor barrier on the exterior. This can be a problem where brick is installed over building paper and vapor permeable sheathing.

Measures Need to be Taken

Two seemingly simple requirements for building enclosures bedevil engineers and architects almost endlessly:

- Keep water out
- Let water out if it gets in

Water can come in several phases: liquid, solid, vapor and adsorbed. The liquid phase as rain and ground water has driven everyone crazy for hundreds of years but can be readily understood - drain everything and remember the humble flashing. The solid phase also drives everyone crazy when we have to shovel it or melt it, but at least most professionals understand the related building problems (ice damming, frost heave, freeze-thaw damage). But the vapor phase is in a class of craziness all by itself. We will conveniently ignore the adsorbed phase and leave it for someone else to deal with. Note that adsorbed water is different than absorbed water.

The fundamental principle of control of water in the liquid form is to drain it out if it gets in – and let us make it perfectly clear – it will get in if you build where it rains or if you put your building in the ground where there is water in the ground. This is easy to understand, logical, with a long historical basis.

The fundamental principle of control of water in the solid form is to not let it get solid and if it does – give it space or if it is solid not let it get liquid and if it does drain it away before it can get solid again. This is a little more difficult to understand, but logical and based on solid research. Examples of this principle include the use of air entrained concrete to control freeze-thaw damage and the use of attic venting to provide cold roof decks to control ice damming.

The fundamental principle of control of water in the vapor form is to keep it out and to let it out if it gets in. Simple right? No chance. It gets complicated because sometimes the best strategies to keep water vapor out also trap water vapor in. This can be a real problem if the assemblies start out wet because of rain or the use of wet materials.

It gets even more complicated because of climate. In general water vapor moves from the warm side of building assemblies to the cold side of building assemblies. This is simple to understand, except we have trouble deciding what side of a wall is the cold or warm side. Logically, this means we need different strategies for different climates. We also have to take into account differences between summer and winter.

Finally, complications arise when materials can store water. This can be both good and bad. A cladding system such as a brick veneer can act as a reservoir after a rainstorm and significantly complicate wall design. Alternatively, wood framing or masonry can act as a hygric buffer absorbing water lessening moisture shocks.

What is required is to define vapor control measures on a more regional climatic basis and to define the vapor control measures more precisely.

Part of the problem is that we struggle with names and terms. We have vapor retarders, we have vapor barriers, we have vapor permeable we have vapor impermeable, etc. What do these terms mean? It depends on whom you ask and whether they are selling something or arguing with a building official. In an attempt to clear up some of the confusion the following definitions are proposed:

Vapor Retarder: The element that is designed and installed in an assembly to retard the movement of water by vapor diffusion.

The unit of measurement typically used in characterizing the water vapor permeance of materials is the "perm." It is further proposed here that there should be several classes of vapor retarders (this is nothing new – it is an extension and modification of the Canadian General Standards Board approach that specifies Type I and Type II vapor retarders – the numbers here are a little different however):

Class I Vapor Retarder:	0.1 perm or less
Class II Vapor Retarder:	1.0 perm or less and greater than 0.1 perm
Class III Vapor Retarder	10 perm or less and greater than 1.0 perm
Test Procedure for vapor retarders:	ASTM E-96 Test Method A (the desiccant method or dry cup method)

Finally, a vapor barrier is defined as:

Vapor Barrier: A Class I vapor retarder.

The current International Building Code (and its derivative codes) defines a vapor retarder as 1.0 perm or less (using the same test procedure). In other words the current code definition of a vapor retarder is equivalent to the definition of a Class II Vapor Retarder proposed by the author.

Continuing in the spirit of finally defining terms that are tossed around in the enclosure business. It is also proposed that materials be separated into four general classes based on their permeance:

Vapor impermeable	0.1 perm or less
Vapor semi-impermeable	1.0 perm or less and greater than 0.1 perm
Vapor semi-permeable:	10 perms or less and greater than 1.0 perm
Vapor permeable:	greater than 10 perms

Recommendations for Building Enclosures

The following building assembly recommendations are climatically based and are sensitive to cladding type (brick or stone veneer, stucco) and structure (concrete block, steel or wood frame, precast concrete).

The recommendations apply to residential, business, assembly, and educational and mercantile occupancies. The recommendations do not apply to special use enclosures such as spas, pool buildings, museums, hospitals, data processing centers or other engineered enclosures such as factory, storage or utility enclosures.

The recommendations are based on the following principles:

- Avoidance of using vapor barriers where vapor retarders will provide satisfactory performance. Avoidance of using vapor retarders where vapor permeable materials will provide satisfactory performance. Thereby encouraging drying mechanisms over wetting prevention mechanisms.

- Avoidance of the installation of vapor barriers on both sides of assemblies – i.e. "double vapor barriers" in order to facilitate assembly drying in at least one direction.

- Avoidance of the installation of vapor barriers such as polyethylene vapor barriers, foil faced batt insulation and reflective radiant barrier foil insulation on the interior of air-conditioned assemblies – a practice that has been linked with moldy buildings.

- Avoidance of the installation of vinyl wall coverings on the inside of air-conditioned assemblies – a practice that has been linked with moldy buildings.

- Enclosures are ventilated meeting ASHRAE Standard 62.2 or 62.1.

Each of the recommended building assemblies were evaluated using dynamic hygrothermal modeling. The moisture content of building materials that comprise the building assemblies all remained below the equilibrium moisture content of the materials as specified in ASHRAE 160 P under this evaluation approach. Interior air conditions and exterior air conditions as specified by ASHRAE 160 P were used. WUFI was used as the modeling program.

More significantly, each of the recommended building assemblies have been found by the author to provide satisfactory performance under the limitations noted. Satisfactory performance is defined as no moisture problems reported or observed over at least a 10-year period.

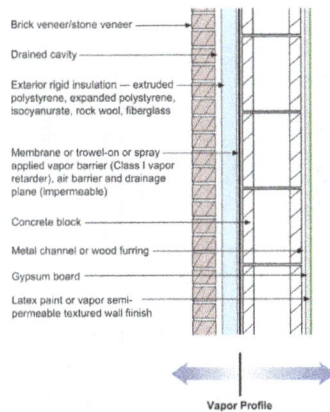

Figure A: Concrete block with exterior
insulation and brick or stone veneer

Applicability – All hygro-thermal regions.

This is arguably the most durable wall assembly available to architects and engineers. It is constructed from non-water sensitive materials and due to the block construction has a large moisture storage (or hygric buffer) capacity. It can be constructed virtually anywhere. In cold climates condensation is limited on the interior side of the vapor barrier as a result of installing all of the thermal insulation on the exterior side of the vapor barrier (which is also the drainage plane and air barrier in this assembly). In hot climates any moisture that condenses on the exterior side of the vapor barrier will be drained to the exterior since the vapor barrier is also a drainage plane. This wall assembly will dry from the vapor barrier inwards and will dry from the vapor barrier outwards.

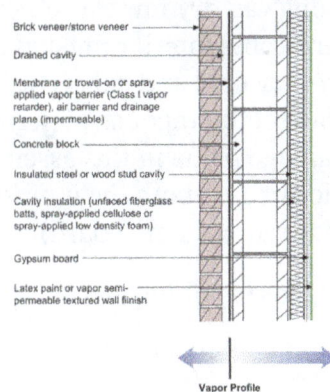

Figure B: Concrete Block With Interior Frame
Wall Cavity Insulation and Brick or Stone Veneer

Applicability – Limited to mixed-humid, hot-humid, mixed-dry, hot-dry and marine regions – should not be used in cold, very cold, and subarctic/arctic regions.

This wall assembly has all of the thermal insulation installed to the interior of the vapor barrier and therefore should not be used in cold regions or colder. It is also a durable assembly due to the block construction and the associated moisture storage (hygric buffer) capacity. The wall assembly does contain water sensitive cavity insulation (except where spray foam is used) and it is important that this assembly can dry inwards – therefore vapor semi impermeable interior finishes such as vinyl wall coverings should be avoided. In this wall assembly the vapor barrier is also the drainage plane and air barrier.

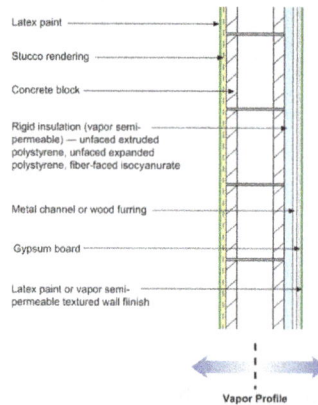

Figure C: Concrete block with interior rigid insulation and stucco

Applicability – All hygro-thermal regions*

This assembly has all of the thermal insulation installed on the interior of the concrete block construction but differs from figure B since it does not have a vapor barrier on the exterior. The assembly also does not have a vapor barrier on the interior of the assembly. It has a large moisture storage (hygric buffer) capacity due to the block construction. The rigid insulation installed on the interior should ideally be non-moisture sensitive and allow the wall to dry inwards – hence the recommended use of vapor semi permeable foam sheathing. Note that foam sheathing faced with aluminum foil or polypropylene skins would also be acceptable provided only non-moisture sensitive materials are used at the masonry block to insulation interface. It is important that this assembly inboard of the foam sheathing can dry inwards except in very cold and subarctic/arctic regions – therefore vapor semi impermeable interior finishes such as vinyl wall coverings should be avoided in assemblies – except in very cold and subarctic/arctic regions. Vapor impermeable foam sheathings should be used in place of the vapor semi permeable foam sheathings in very cold and subarctic/arctic regions. The drainage plane in this assembly is the latex painted stucco rendering. A Class III vapor retarder is located on both the interior and exterior of the assembly (the latex paint on the stucco and on the interior gypsum board).

* In very cold and subarctic/arctic regions vapor impermeable foam sheathings are recommended

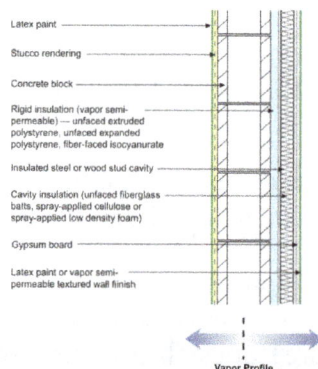

Figure D: Concrete block with interior rigid
insulation/frame wall with cavity insulation and stucco

Applicability – All hygro-thermal regions*

This assembly is a variation of figure C. It also has all of the thermal insulation installed on the interior of the concrete block construction but differs from figure C, due to the addition of a frame wall to the interior of the rigid insulation. This assembly also does not have a vapor barrier on the exterior. The assembly also does not have a vapor barrier on the interior of the assembly. It has a large moisture storage (hygric buffer) capacity due to the block construction. The rigid insulation installed on the interior should ideally be non-moisture sensitive and allow the wall to dry inwards — hence the recommended use of vapor semi permeable foam sheathing. Note that foam sheathing faced with aluminum foil or polypropylene skins would also be acceptable provided only non-moisture sensitive materials are used at the masonry block to insulation interface. It is important that this assembly inboard of the rigid insulation can dry inwards even in very cold and subarctic/arctic regions — therefore vapor semi impermeable interior finished such as vinyl wall coverings should be avoided in assemblies. Vapor impermeable foam sheathings should be used in place of the vapor semi permeable foam sheathings in very cold and subarctic/arctic regions. The drainage plane in this assembly is the latex painted stucco rendering. A Class III vapor retarder is located on both the interior and exterior of the assembly (the latex paint on the stucco and on the interior gypsum board).

* In very cold and sub arctic/arctic regions vapor impermeable foam sheathings are recommended – additionally the thickness of the foam sheathing should be determined by hygro-thermal analysis so that the interior surface of the foam sheathing remains above the dew point temperature of the interior air.

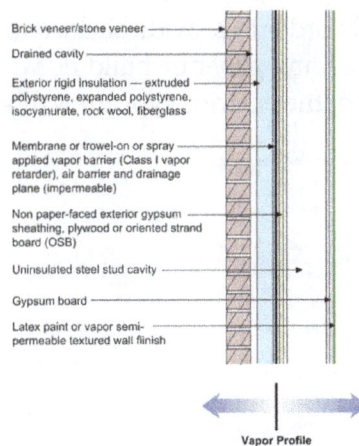

Figure E: Frame wall with exterior insulation and brick of stone veneer

Applicability – All hygro-thermal regions

This wall is a variation of figure A, but without the moisture storage (or hygric buffer) capacity. This wall is also a durable wall assembly. It is constructed from non-water sensitive materials and has a high drying potential inwards due to the frame wall cavity not being insulated. It can also be constructed virtually anywhere. In cold climates condensation is limited on the interior side of the vapor barrier as a result of installing all of the thermal insulation on the exterior side of the vapor barrier (which is also the drainage plane and air barrier in this assembly). In hot climates any moisture that condenses on the exterior side of the vapor barrier will be drained to the exterior since the vapor barrier is also a drainage plane. This wall assembly will dry from the vapor barrier inwards and will dry from the vapor barrier outwards.

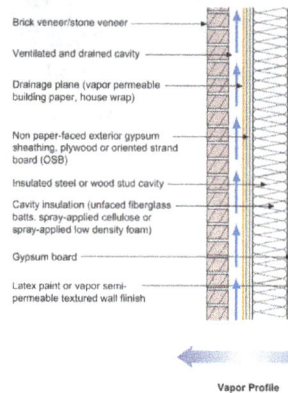

Figure F: Frame Wall With Cavity Insulation and Brick or Stone Veneer

Applicability – Limited to mixed-humid, hot-humid, mixed-dry, hot-dry and marine regions – can be used with hygro-thermal analysis in some areas in cold regions- should not be used in very cold and subarctic/arctic regions

This wall is a flow through assembly – it can dry to both the exterior and the interior. It has a Class III vapor retarder on the interior of the assembly (the latex paint on the gypsum board). It is critical in this wall assembly that the exterior brick veneer (a "reservoir" cladding) be uncoupled from the wall assembly with a ventilated and drained cavity. The cavity behind the brick veneer should be at least 2 inches wide and free from mortar droppings. It must also have air inlets ("weep holes") at its base and air outlets ("weep holes") at its top in order to provide back ventilation of the brick veneer. The drainage plane in this assembly is the building paper or building wrap. The air barrier can be any of the following: the interior gypsum board, the exterior gypsum wallboard or the exterior building wrap.

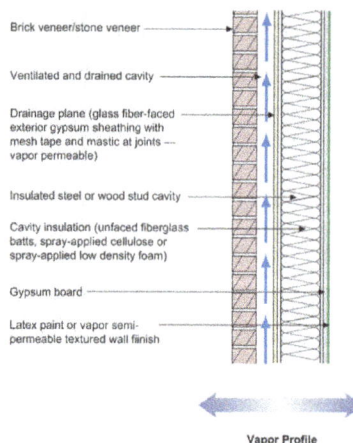

Figure G: Frame Wall With Cavity Insulation and Brick or Stone Veneer

Applicability – Limited to mixed-humid, hot-humid, mixed-dry, hot-dry and marine regions – can be used with hygro-thermal analysis in some areas in cold regions - should not be used in very cold and subarctic/arctic regions

This wall is a variation of figure F. The exterior gypsum sheathing becomes the drainage plane. As in figure F, this wall is a flow through assembly – it can dry to both the exterior and the interior. It has a Class III vapor retarder on the interior of the assembly (the latex paint on the gypsum board). It is also critical in this wall assembly that the exterior brick veneer (a "reservoir" cladding) be uncoupled from the wall assembly with a ventilated and drained cavity. The cavity behind the brick veneer should be at least 2 inches wide and free from mortar droppings. It must also have air inlets ("weep holes") at its base and air outlets ("weep holes") at its top in order to provide back ventilation of the brick veneer. The air barrier in this assembly can be either the interior gypsum board or the exterior gypsum sheathing.

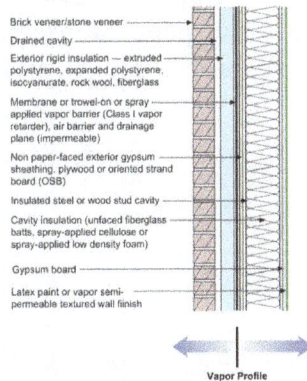

Figure H: Frame Wall With Exterior Rigid InsulationWith Cavity
Insulation and Brick or Stone Veneer

Applicability – All hygro-thermal regions except subarctic/arctic – in cold and very cold regions the thickness of the foam sheathing should be determined by hygro-thermal analysis so that the interior surface of the foam sheathing remains above the dew point temperature of the interior air.

This wall is a variation of figure E. In cold climates condensation is limited on the interior side of the vapor barrier as a result of installing some of the thermal insulation on the exterior side of the vapor barrier (which is also the drainage plane and air barrier in this assembly). In hot climates any moisture that condenses on the exterior side of the vapor barrier will be drained to the exterior since the vapor barrier is also a drainage plane. This wall assembly will dry from the vapor barrier inwards and will dry from the vapor barrier outwards. Since this wall assembly has a vapor barrier that is also a drainage plane it is not necessary to back vent the brick veneer reservoir cladding as in figures F and figure G. Moisture driven inwards out of the brick veneer will condense on the vapor barrier/drainage plane and be drained outwards.

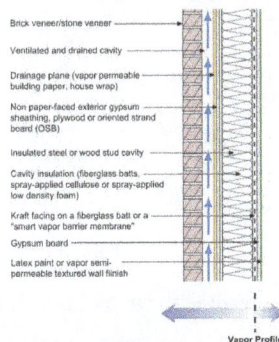

Figure I: Frame Wall With Cavity Insulation and Brick or Stone Veneer
With Interior Vapor Retarder

Applicability – Limited to cold and very cold regions

This wall is a variation of figure F except it has a Class II vapor retarder on the interior limiting its inward drying potential – but not eliminating it. It still considered a flow through assembly – it can dry to both the exterior and the interior. It is critical in this wall assembly- as figure F and figure G- that the exterior brick veneer (a "reservoir" cladding) be uncoupled from the wall assembly with a ventilated and drained cavity. The cavity behind the brick veneer should be at least 2 inches wide and free from mortar droppings. It must also have air inlets ("weep holes") at its base and air outlets ("weep holes") at its top in order to provide back ventilation of the brick veneer. The drainage plane in this assembly is the building paper or building wrap. The air barrier can be any of the following: the interior gypsum board, the exterior gypsum board or the exterior building wrap.

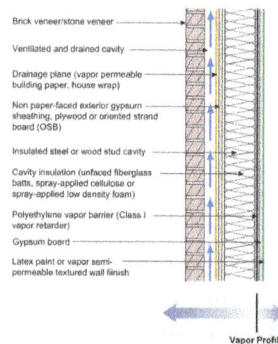

Figure J: Frame wall with cavity insulation and brick or stone
veneer with interior vapor barrier

This wall is a further variation of figure F, but now it has a Class I vapor retarder on the interior (a "vapor barrier") completely eliminating any inward drying potential. It is considered the "classic" cold climate wall assembly as in figure F, figure G and figure H. It is critical in this wall assembly that the exterior brick veneer (a "reservoir" cladding) be uncoupled from the wall assembly with a ventilated and drained cavity. The cavity behind the brick veneer should be at least 2 inches wide and free from mortar droppings. It must also have air inlets at its base and air outlets at its top in order to provide back ventilation of the brick veneer. The drainage plane in this assembly is the building paper or building wrap. The air barrier can be any of the following: the interior polyethylene vapor barrier, the interior gypsum board, the exterior gypsum board or the exterior building wrap.

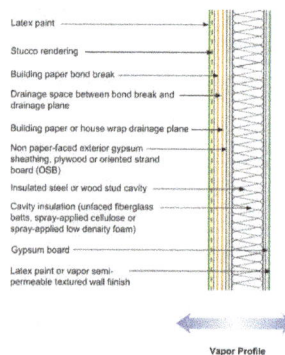

Figure K: Frame wall with cavity insulation and stucco

Applicability – Limited to mixed-humid, hot-humid, mixed-dry, and hot-dry regions should not be used in marine, cold, very cold, and subarctic/arctic regions.

This wall is also a flow through assembly similar to figure F, but without the brick veneer – it has a stucco cladding. It can dry to both the exterior and the interior. It has a Class III vapor retarder on the interior of the assembly (the latex paint on the gypsum board). It is critical in this wall assembly that a drainage space be provided between the stucco rendering and the drainage plane. This can be accomplished by installing a bond break (a layer of tar paper) between the drainage plane and the stucco. A spacer mat can also be used to increase drainability. Alternatively, a textured or profiled drainage plane (building wrap) can be used. The drainage plane in this assembly is the building paper or building wrap. The air barrier can be any of the following: the interior gypsum board, the exterior stucco rendering, the exterior sheathing or the exterior building wrap.

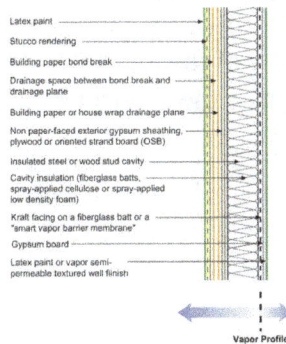

Figure L: Frame Wall With Cavity Insulation and Stucco
With Interior Vapor Retarder

Applicability – Limited to marine, cold and very cold regions

This wall is a variation of figure F and figure K, except it has a Class II vapor retarder on the interior limiting its inward drying potential – but not eliminating it. It still considered a flow through assembly – it can dry to both the exterior and the interior. It is critical in this wall assembly – as in Figure K – that a drainage space be provided between the stucco rendering and the drainage plane. This can be accomplished by installing a bond break (a layer of tar paper) between the drainage plane and the stucco. A spacer mat can also be used to increase drainability. Alternatively, a textured or profiled drainage plane (building wrap) can be used. The drainage plane in this assembly is the building paper or building wrap. The air barrier can be any of the following: the interior gypsum board, the exterior stucco rendering, the exterior sheathing or the exterior building wrap.

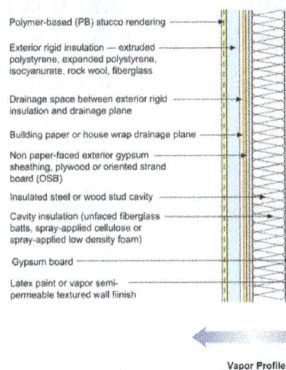

Figure M: Frame wall with exterior rigid insulation
with cavity insulation and stucco

Applicability – All hygro-thermal regions except subarctic/arctic – in cold and very cold regions the thickness of the foam sheathing should be determined by hygro-thermal analysis so that the interior surface of the foam sheathing remains above the dew point temperature of the interior air

This is a water managed exterior insulation finish system (EIFS). Unlike "face-sealed" EIFS this wall has a drainage plane inboard of the exterior stucco skin that is drained to the exterior. It is also a flow through assembly similar to Figure F. It can dry to both the exterior and the interior. It has a Class III vapor retarder on the interior of the assembly (the latex paint on the gypsum board). It is critical in this wall assembly that a drainage space be provided between the exterior rigid insulation and the drainage plane. This can be accomplished by installing a spacer mat or by providing drainage channels in the back of the rigid insulation. Alternatively, a textured or profiled drainage plane (building wrap) can be used. The drainage plane in this assembly is the building paper or building wrap. The air barrier can be any of the following: the interior gypsum board, the exterior stucco rendering, the exterior sheathing or the exterior building wrap.

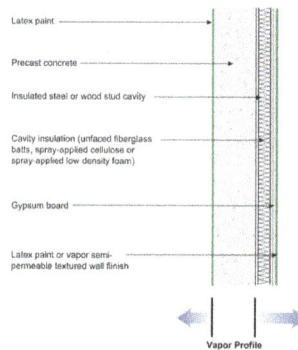

Figure N: Precast concrete with interior frame wall cavity insulatioin

*Applicability – Limited to mixed-humid, hot-humid, mixed-dry, hot-dry and marine regions – should not be used in cold, very cold, and subarctic/arctic regions.

The vapor barrier in this assembly is the precast concrete itself. Therefore this wall assembly has all of the thermal insulation installed to the interior of the vapor barrier. Of particular concern is the fact that the thermal insulation is air permeable (except where spray foam is used). Therefore this wall assembly should not be used in cold regions or colder. It has a small moisture storage (hygric buffer) capacity due to the precast concrete construction. The wall assembly does contain water sensitive cavity insulation (except where spray foam is used) and it is important that this assembly can dry inwards – therefore vapor semi impermeable interior finishes such as vinyl wall coverings should be avoided. In this wall assembly the precast concrete is also the drainage plane and air barrier.

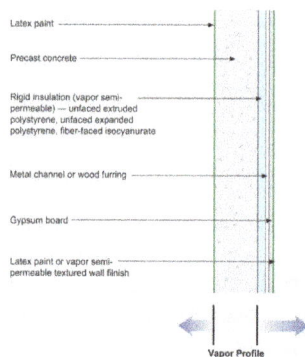

Figure O: Precast Concrete With Interior Rigid Insulation

This assembly has all of the thermal insulation installed on the interior of the precast concrete. The assembly also does not have a vapor barrier on the interior of the assembly. It has a small moistuure storage (hygric buffer) capacity due to the precast concrete construction. The rigid insulation installed on the interior should ideally be non-moisture sensitive and allow the wall to dry inwards — hence the recommended use of vapor semi permeable foam sheathing. Note that foam sheathing faced with aluminum foil or polypropylene skins would also be acceptable provided only non-moisture sensitive materials are used at the concrete to insulation interface. It is important that this assembly inboard of the foam sheathing can dry inwards except in very cold subarctic/arctic regions — therefore vapor semi impermeable interior finishes such as vinyl wall coverings should be avoided in assemblies — except in very cold and subarctic/arctic regions. Vapor impermeable foam sheathings should be used in place of the vapor semi permeable foam sheathings in very cold and subarctic/arctic regions. The drainage plane in this assembly is the latex painted precast concrete. A Class III vapor retarder is located on both the interior and exterior of the assembly (the latex paint on the stucco and on the interior gypsum board).

* In very cold and subarctic/arctic regions vapor impermeable foam sheathings are recommended

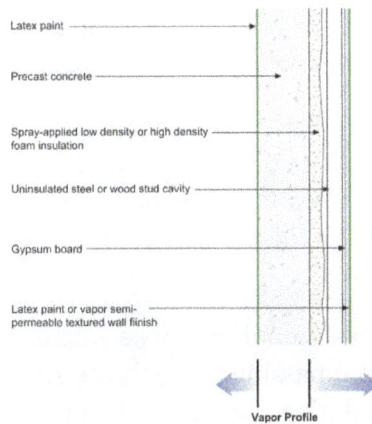

Figure P: Precast Concrete With Interior
Spray Applied Foam Insulation

Applicability – All hygro-thermal regions*

This assembly has all of the thermal insulation installed on the interior of the precast concrete. The assembly also does not have a vapor barrier on the interior of the assembly. It has a small moisture storage (hygric buffer) capacity due to the precast concrete construction. The spray foam insulation installed on the interior of the precast concrete is non-moisture sensitive and allows the wall to dry inwards. It is important that this assembly can dry inwards except in very cold and subarctic/arctic regions – therefore vapor semi impermeable interior finishes such as vinyl wall coverings should be avoided in assemblies – except in very cold and subarctic/arctic regions. High-density spray foam, due to its vapor semi impermeable characteristics should be used in place of low-density foam in very cold and subarctic/arctic regions. The drainage plane in this assembly is the latex painted precast concrete. A Class III vapor retarder is located on both the interior and exterior of the assembly (the latex paint on the stucco and on the interior gypsum board.

* In very cold and subarctic/arctic regions high-density spray foam (vapor semi impermeable) is recommended.

Housewrap

Contractors use house wrap to cover the wood frame before the application of stone, wood, brick, or vinyl siding. House wrap helps keep moisture away from the interior of the house and protects the surface of the exterior walls.

House wrap is made with water-resistant materials such as polymers, nylon, or fiberglass, but it can also be made of polyethylene. It is usually a yellow or white paper-like sheet that has a glossy sheen. It comes in wide rolls with a width measuring up to 9 feet. As the name suggests, a house wrap functions more like gift-wrap protecting and sealing the item inside by covering all edges and functioning as a boxing cover around the house.

1. Virtually all houses have wood in their structure, moisture during the framing of the house is a dilemma you will face from time to time especially if you don't water-proof your house. When moisture seeps into it, it could rot, encourage mold growth, and cause health hazards to the occupants. House wrap can help prevent these things by protecting the structure and surface of your house from moisture damage. The plastic membrane of the house wrap acts as a barrier and prevents water seepage. As it keeps the structure dry inside, you will not have to worry about these problems.

2. Aside from shielding the house' structure from damaging buildup of moisture, a house wrap also serves as a good insulating material. The plastic-like, non-airtight feature of a house wrap limits air flow and reduces heat exchange from the interior and exterior of your home. This insulating effect, in turn, helps ease the burden on the part of your home conditioning system, helps you save energy, and lower your utility bills.

3. Not only does house wrap serve as a multi-purpose weather barrier, it also acts as a drainage plane. House wrap blocks water from passing through the structure but lets water vapor seep inside. This, then, allows moist humid air to break out from the inside while keeping liquid water off the interior of the home. Instead, it directs the rain water away from the house structure and straight to the ground.

Aside from these three main functions, another benefit of using house wrap is in terms of its durability. Although made of a thin sheet, house wrap can withstand normal wear and tear. If installed properly, it could last for a long time without showing signs of deterioration. However, according to Federal Emergency Management Administration or FEMA, compared to building paper, the latter can endure extreme hot or extreme cold climates and high winds better. Therefore, house wrap may not work well on places with these kinds of climate conditions.

Overall, house wrap is a great investment you should consider when constructing a new home. Though it has a number of advantages, installing it the wrong way can prove very hazardous rather than useful. FEMA has instructed that the sheets should be installed in horizontal rows with its top and bottom overlapping and seams covered with protective tape.

There are specialists who can do the job for you if you believe you are not up for the challenge. Proper installation is the key to making your house wrap work for you. If you are unsure how to go about installing house wrap, check with your local hardware store for some advice. Contractors in your area may also be able to come out to see the job and offer their expertise.

Basement Waterproofing

The basement waterproofing systems prevent water from leaking into the basement and damaging the foundation and wood. Basement waterproofing systems are needed whenever a basement, cellar, or other room is built at ground level or below. They are particularly important in areas where it is likely that ground water will raise the water table.

Without proper basement waterproofing systems, the basement can become cracked, and water pressure can result in serious damage to walls, as well as the growth mold and wood rot. Building the right kind of waterproofing system depends very much upon the environment in which the house is built.

Interior Sealant

Using interior sealants as basement waterproofing systems is a temporary measure, to be used during winter months to prevent snow and frost from raising the water table.

These sealants are often found in chemical spray form, and can be directly applied to walls and floors. Interior sealants can also prevent humidity and condensation within the basement. They can be absorbed by woods and porous building materials, causing cracks in masonry, damage to concrete, and wood rot. Masonry is also protected against spalling.

The most common use of these sealants is to prevent humidity inside the house from affecting the walls of the basement. For more secure measures against ground water, you will need to have either an exterior sealant or a drainage system.

Exterior Sealant

Exterior basement waterproofing systems stop ground water from reaching the basement walls as well as prevent mold and other damage which can occur in wet basement areas.

Waterproofing an exterior is the recognized IBC method to prevent damage caused by water. Exterior sealants were once only asphalt-based damp proofing, but now the most common kind of exterior basement waterproofing systems use a polymer base. This kind of material will last for the life of the building.

Other exterior sealants will probably erode in 20 years. You will not usually have to add anything to an exterior sealant, which should applied during the construction of the house, particularly in areas where flooding, heavy rain, or hurricanes are likely to occur.

Water Drainage

Drainage can be used to mitigate basement water and is often considered to be another form of basement waterproofing. Water drainage functions by drawing water away from the foundation, and forcing it into a drain, or through a pump system. Pump kids are available which can be installed in DIY form or by professional plumbers.

Water drainage basement waterproofing systems often need to be run on an isolated electric system in case of power shortage, especially during periods of storms. Sump pumps should be placed in a pit and sealed with a lid in order to keep the water away from the electricity. Doing so also prevents humidity in the pump from entering the atmosphere of the basement. Keeping the lid airtight ensures that poisonous gasses won't seep into the house.

Methods Used in Basement Waterproofing

Two methods used in basement waterproofing is the Positive(application on the same side of hydro-static pressure) and the Negative(application on the opposite side of water pressure) waterproofing.

- Positive Waterproofing: In this method, the surrounding perimeters to the building structure is dug – about a meter deep(depends) and layers of waterproofing solution is applied. This is the best suggested way to protect since the major cause of problem is water pressure from the outside during monsoons.

- Negative Waterproofing: This method must be used in extreme conditions when there is no other way of performing the task from the outside. Cautions must be taken that the materials used are not permeable and do not form honeycomb structures. This application is not very reliable because when the water seeps in, the proofing layer is not always flexible enough to adhere that much pressure. It has been a trend by many contractors to use this method because of the ease and faster application. But this must be avoided and Positive (outside) applications must be undertaken.

- Already Constructed basements can also apply a layer of Kota – Patthar/Stone tiles with Injection Grouting method.

A wet basement not only prevents you from enjoying additional space in your house but also can turn your basement into a giant petri dish perfect for growing unhealthy molds and fungi.

A Dry Basement is a Useable Basement

Installing a drain system is filthy, backbreaking work, but it's not complicated. With a little instruction from our drain tile experts, you can do a first-class job. And DIY pays off big: Pros charge $5,000 to $8,000 for a typical job (120 linear feet of drain tile). You can install yours for less than $1,500 in materials and tool rentals.

Rite-Way Waterproofing has been in the business of installing waterproofing basement drain systems since 1965. The pros there know what works, and they'd better, because the company has installed 100,000 systems, each with a lifetime guarantee. They also know what doesn't work—half of Rite-Way's current jobs involve replacing failed systems installed by other contractors.

A drain tile system creates a perfect pathway for dangerous radon gas to escape. If you've never tested for radon, it's smart to do so before you install a drain system. That way, you can plan for a radon mitigation system as well.

Before you Get Started

It's always best to stop water from entering your basement in the first place, so before you run to the rental center for your jackhammer, be sure to address the exterior issues. The grade next to the house should slope down away from the building by least 6 in. for the first 10 ft. Consider installing gutters, or make sure the existing gutters are working properly. And check that your irrigation system isn't adding to the problem by spraying water right up against the side of the house.

If your waterproofing basement is finished, with stud walls and insulation covering the foundation walls, you can still install a drain system. When you break out the concrete, leave small sections of floor intact so the wall doesn't drop down. A 4×4 in. section every 6 ft. is enough to support the wall. If there are obstacles along the wall (like a furnace), plan to tunnel under them. You'll find most of the materials you'll need at a home center. Order the rock from a landscape supplier. You'll also need a pickup to haul the dirt to the landfill.

Always check with your local building official. Explain your project, and see if any permits or inspections are required in your area. Sometimes, a building official who has been around for a while may have information on how your house was built or what issues you may run into in your area.

Control the Dust

Get Ready for Dust

Instead of just covering your stuff with sheets of plastic, isolate your work area with a wall of plastic sheeting. Make sure to fill in the spaces between joists.

There's no getting around it: Busting up concrete is a dirty job. Shut down your furnace or central air conditioning while you're working, and cover all return air vents until you're finished cleaning up. Instead of covering your furnishings with plastic, move everything out of the area and drape plastic from the ceiling to create an isolated work space. If you have an unfinished ceiling, be sure you run the plastic up into every joist space. Set a fan in the window to exhaust the heavy dust while you run the jackhammer. And wear a dust mask and hearing protection.

Bust up the Floor

Bust up the Floor.

Remove 16 in. to 18 in. of concrete along the wall with a rented electric jackhammer. Start by chipping in a straight line along the entire length of the wall, then come back and bust it into manageable chunks.

The pros use electric jackhammers, because the air that runs pneumatic jackhammers kicks up a lot more dust. You can get one from a rental center for under $100 a day. Start by hammering a line about 16 in. to 18 in. away from the wall.

Once the perimeter is done, come back and break the row of concrete into manageable chunks. Each section will break free easier if it has room to pull away, so remove the sections as you go. If you're working alone, make the most of your rental time; just set the chunks aside until you're done with the hammer. Don't forget to bust up a larger area for your sump basin.

Dig the Trench

Dig a trench along the wall.

Dig the trench as deep as the bottom of the footing. Instead of lugging pails of soil up the stairs, buy buckets that will fit through your basement windows.

Once the concrete is removed, dig down to the bottom of the footing but not below. If you compromise the soil under the footing, you could end up with cracks in your wall, or worse.

Five-gallon buckets are OK for hauling out debris, but the guys at Rite-Way use rubber feed buckets (typically used for farm animals) because they fit through small basement windows and are less likely to bang up trim. You can get them at farm supply retailers for under $20. And when it's time to haul the debris away, you may find that the landfill considers it to be "clean fill." You may be able to dump it for free.

Install the Basin

It's best to locate your basin in an unfinished area of the basement so you can have easy access to the sump pump. If you never plan on finishing the basement, locate the basin in the same area where you want the water to drain out of the house so you don't have as much plastic pipe to install. Dig the hole so the top of the basin will sit flush with the finished concrete.

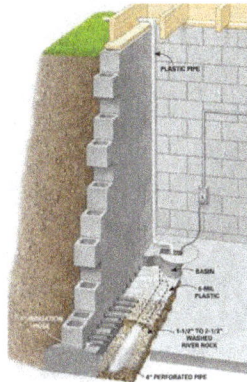

Many basins come with flat "knockout" areas meant to make cutting the hole easier. Don't assume the location of these knockouts will work for your system. Because the pipe will be slightly sloped down toward the basin, the longer the drain is, the lower the pipe will be when it reaches the basin. You never want standing water in your drainpipes, so make sure to choose a model that is deep enough. Our experts typically use 30-in.-deep basins. They use 36-in.-deep models for systems longer than 120 ft., and they install two basins if the drain is longer than 180 ft.

Set the basin in place, and then mark the locations for the holes where the pipes will meet the basin. Keep in mind there will be a thin layer of rock (one layer thick) under the pipe near the basin. Cut the holes using a reciprocating saw, jigsaw or hole saw. The holes don't have to be perfect.

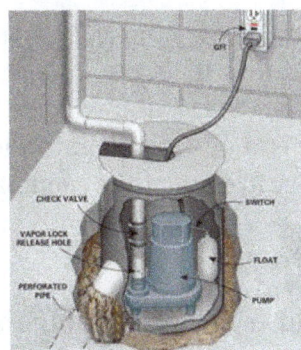

Don't haul out all the dirt right away; you'll need some to fill in around the basin. Once it's permanently in place, fill in around it, tamping the dirt with a 2×4 as you go.

Caution: Never drill holes in the bottom of the basin. If you have a high water table, water could come up from the bottom, and your pump will run nonstop, attempting to dry out the neighborhood.

Drill Holes in Block

If your waterproofing basement walls are made from concrete block, drill 1-in. holes in each block core and into each mortar joint. This will allow the water that collects in the cores and between the blocks to flow into the drain. Drill the holes as close to the footings as you can. You may find that the bottom blocks are filled with concrete. In that case, you'll have to remove any existing walls and install foundation wrap. Cut down on dust by laying a shop vacuum hose next to the hole as you drill.

Install the Pipe

Lay in the pipe

Lay the pipe with the holes face down. Add or remove rock to slope the pipe so water flows to the sump basin. There's no need to cement the pipe connections

Before you lay the pipe in the trench, shovel in a bottom layer of 1-1/2-in. to 2-in. washed river rock (a layer of smaller rock can become clogged with minerals and sediments). The pipe should slope toward the basin at least 1/4 in. for every 10 ft. Rake the rock around to achieve this pitch.

Lay your irrigation pipe on top of the rock. Don't use ordinary flexible drainpipe because it clogs easily. The pros prefer to lay down a 4-in.-diameter Schedule 10 perforated pipe. Buy the kind of pipe with rows of 1/2-in. perforation holes only on one side, not all around the pipe.

Lay the pipe with the holes facing down, so the minerals and sediment in the water can flow down around the pipe and settle into the ground. This way, the water that does rise up into the pipes from underneath will be relatively clean. Clean water will add years to the life of the whole system. Start at the basin, and push the male end of the pipe into the basin about 4 in. Use PVC or ABS elbows at the corners. It's not necessary to cement the sections together.

Divert the Water into the Drain

Drill 1-in. holes in each block core and each mortar joint. Then insert sections of 1-in. irrigation hose from the holes into the gravel to carry away the water.

Once your pipe is installed, it's time to install the 1-in. irrigation hose that will carry the water from the blocks to the trench. Softer hose, like garden hose, can get crushed flat by the new concrete, so stick with irrigation hose. Cut the hose with a hacksaw or reciprocating saw. Make sure each section of hose runs several inches past the footing.

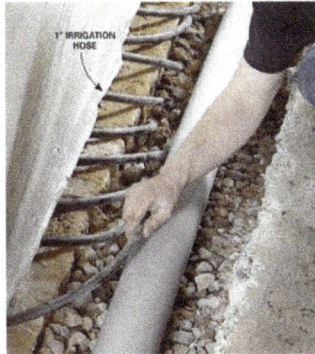

Install hose in the block walls

If you have a Solid Concrete Foundation

If you have poured concrete walls rather than block, you'll need to install a foundation wrap to let the water into the drain as shown below. Foundation wrap is made from tough plastic and consists of rows of dimples that allow water to flow behind it. Platon by CertainTeed is one example. Home centers can order it if they don't stock it.

Cut the sheets into strips with a utility knife. Bend the strips at 90 degrees and let the bottom half run past the footing. The length of the wrap that's up against the wall depends on your situation. At a bare minimum, run the wrap up 4 in. above the top of the concrete on a poured wall, or up 4 in. past the holes you drilled in the block wall. If you're working along stud walls, try to tuck the wrap behind the bottom plate.

Cover it Back up

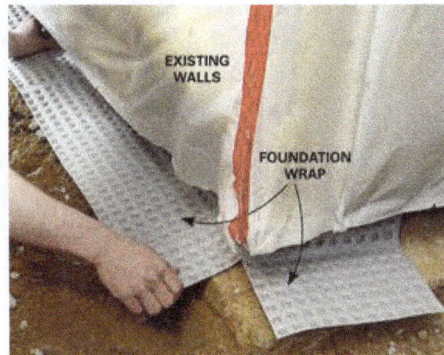

Cap off the trench with concrete

Lay plastic over the rock and cover it with concrete. Smooth out the concrete with a float. Wait 20 minutes, and work it smooth with a steel trowel.

Once the hoses are in or the foundation wrap is in place, it's time to cover it up. Fill in the trench with river rock up to the bottom of the existing slab, and then cover the rock with at least a 6-mil thickness of plastic for a vapor barrier.

To cut down on dust, mix your concrete outside. Bagged concrete mix for slabs and sidewalks works just fine. Slide a 3-ft. section of 2×4 along the floor to "screed" the new concrete flush with the floor, and then smooth it out with a hand float. Wait 20 minutes, then smooth it with a finishing trowel. Use the float to completely fill the gap under any existing walls.

Hook up the Pump

Pros prefer submersible pumps that have a vertical float switch on them because they're more reliable than pedestal or float switch pumps. Install a 6- to 8-in. section of pipe on the pump, then a check valve. Make sure the check valve doesn't interfere with the pump switch. Above the check valve, attach another section of pipe long enough to reach above the top of the basin.

Drill a 1/4-in. to 3/8-in. vapor lock release hole in the section of pipe that's just below the check valve. This allows the pump to get up to speed before trying to force open the check valve, which may have many gallons of water pressing down on it. Angle the hole so water sprays down while the pump is working. It's best if you have a dedicated outlet for your pump. Extension cords get unplugged, and other appliances hooked up to the same circuit could trip a breaker.

The pipe that exits the basement needs to be located in an area that slopes away from the house. If that means running a pipe back across the basement, consider burying the waste pipe in the trench and having it come back up where you want it. If your pipe is going to discharge above grade and you live in a cold climate, run the pipe no more than 8 in. past the siding. This will keep it from freezing up in the winter.

With few exceptions, basement drain water cannot be dumped into city sewer systems. Most systems can be drained into storm sewers as long as they're above grade when they do. Ask your building official what the rules are in your area.

Required Tools for this Project

Have the necessary tools for this DIY project lined up before you start—you'll save time and frustration.

- Bucket
- Chalk line
- Cold chisel
- Dust mask
- Extension cord
- Hacksaw
- Hammer drill
- Jigsaw
- Knee pads
- Knockdown knife
- Reciprocating saw

- Safety glasses

- Shop vacuum

- Spade

- Tape measure

- Trowel

- Wheelbarrow

Flashing

Flashing (weatherproofing) is a product used to provide a weatherproof seal around penetrations in a building's roof or walls. It consists of a thin layer of waterproof material that keeps water from getting into places it does not belong. It is used around roof penetrations such as the edges of skylights, chimneys, vent stacks and vent fans as well as at intersections of roofing surfaces such as roof valleys. It is also used around penetrations in vertical surfaces such as around window and door sills. As water lands on a building's exterior, flashing directs it over and past crevices, cracks and gaps so that it does not come in contact with vulnerable building materials.

Flashing must be installed so no seams face upward. It is usually layered with other building materials to provide a clear flow of water past any openings in the building. For instance, the upper edge of flashing is protected by house wrap or tarpaper, not installed over it.

Types

Below are some common types of flashing. Most are defined by the application in which they will be used. There are many different shapes and configurations of flashing available to accommodate its various uses.

- Chimney flashing is applied around the base of a chimney in several parts.

- Continuous flashing protects the joint between a vertical wall and a sloped roof.

- Drip edges prevent water from seeping under roofing along the edges of rakes and eaves.

- Step flashing steps up a roof to protect where the roof meets the side walls of dormers.

- Valley flashing protects the valleys where two roof planes meet.

- Vent pipe flashing fits over flues and pipes. It is cone-shaped with a flange at the base.

- Window flashing - This is the basic thin sheet of metal that can be easily bent and molded to fit a space.

- Tape—Newest form of flashing and comes as a self-adhering flexible membrane.

Materials

Many materials can be used for flashing. As long as it won't degrade from contact with incompatible materials it should work. Flashing is typically produced as a sheet of aluminum, copper, PVC, steel, or lead, but may also take the form of a flexible rubberized sheet.

- Aluminum—Easy to form, durable and relatively inexpensive

- Bituminous Ffashing tape—Tar like material with a sticky backing

- Copper and lead-coated copper—Harder to form than aluminum but very durable

- Lead—Very soft and easy to bend. Very durable

- Galvanized steel—Inexpensive but not as durable as others

- PVC—Easy to work with and inert

Tar Paper

Tar paper is made by soaking a porous paper made from cotton rag scraps with thinned liquid asphalt. Asphalt, of course, is one of the final products that comes out of a catalytic convertor that's used to refine crude oil.

The paper comes in different weights. The most common weights are 15-pound and 30-pound tar paper. The 30-pound tar paper is heavier and has much more asphalt in it.

While impossible to know unless you do an expensive analysis, the asphalt used to make modern tar paper contains much more oxygen than it should.

The tar paper made between 1900 and 1980 probably will last far longer than the tar paper made today.

Fiberglass Mats

Much of today's tar paper is made using multiple fiberglass mats as the cotton rag industry has declined significantly over the past few decades. This is just part of the reason why fiberglass shingles were introduced. There simply was a shortage of cotton scrap to make the mats that are the foundation of shingles.

The big reason fiberglass pushed aside cotton-fiber mats was they could run the giant mills three times faster than if they used cotton. This means more profits for the asphalt shingle manufacturers.

UV Damage

Since felt paper gets covered with something not long after it's installed, there's little chance it will degrade. The ultraviolet rays from the sun attack the exposed asphalt and cause it to oxidize and cross link with adjacent asphalt molecules. This cross linking makes the asphalt brittle.

Time Tested

Using tar paper to protect wood sheathing and wood framing members on houses, room additions or outdoor sheds is a fantastic idea. This time-tested product is affordable, it's easy to work with and it's readily available.

It's all about shedding water. There are pre-printed lines on the tar paper that help show you where to end the overlap. Usually 2 inches is plenty on a horizontal seam. If you have a vertical seam where one piece ends and another starts, make the overlap at least 6 inches.

Overlap Top of Foundation

Another great installation tip is to make sure the first strip of tar paper is installed so it overlaps the top of the foundation at least an inch. You want any water that does get behind the siding to run down and never be allowed to get near any wood. Many homeowners and builders fail to create this mission-critical overlap at the foundation.

New Weather Barriers

The newer weather barriers made from synthetic fabrics are great products. I've used them as well as tar paper. Some of the new products come in tall 10-foot-wide rolls that allow you to cover a typical one-story house with only one vertical overlap seam.

It's not uncommon to have a roll that's over 100 feet long, if not longer.

You surely can't do that with tar paper as it usually only comes in rolls 3-feet wide. This means you'll have at least four horizontal overlap seams in a typical single-story home.

More Labor

Tar paper will take more labor to install than the newer wider synthetic weather barriers, but if you're doing the work, it costs you just your time. You just need to do the math to see what material will save you money.

Drainage Channels

Some of the newer weather barriers have great drainage channels built into them. These channels help direct water quickly down and away from the exterior siding material. They also promote quick drying allowing air to get behind any siding. This is a good thing.

Tar paper does not offer this. Siding applied directly over tar paper creates a sandwich effect and can trap water between the siding material and the tar paper.

If you want vertical drainage with tar paper, you have to add treated lumber strips on top of the tar paper. This is a time-intensive process and requires all sorts of skill.

References

- Building-envelope, encyclopedia: energyeducation.ca, Retrieved 26 March 2018

- Breathability, how-to-deal-with-damp: Retrieved 20 June 2018

- What-is-damp-proofing, damp-proofing: rentokil.co.uk, Retrieved 12 April 2018

- Moisture-damage-465: comfortsolutions.ie, Retrieved 22 March 2018

- Planning-24: fromplattoplace.com, Retrieved 25 April 2018

- Types-of-waterproofing-basement-waterproofing, home-improvement: broomberg.in, Retrieved 16 July 2018

- Tar-paper-facts-and-tips: askthebuilder.com, Retrieved 18 June 2018

Permissions

We would like to thank the editorial team for lending their expertise to make the book truly unique. They have played a crucial role in the development of this book. Without their invaluable contributions this book wouldn't have been possible. They have made vital efforts to compile up to date information on the varied aspects of this subject to make this book a valuable addition to the collection of many professionals and students.

This book was conceptualized with the vision of imparting up-to-date and integrated information in this field. To ensure the same, a matchless editorial board was set up. Every individual on the board went through rigorous rounds of assessment to prove their worth. After which they invested a large part of their time researching and compiling the most relevant data for our readers.

The editorial board has been involved in producing this book since its inception. They have spent rigorous hours researching and exploring the diverse topics which have resulted in the successful publishing of this book. They have passed on their knowledge of decades through this book. To expedite this challenging task, the publisher supported the team at every step. A small team of assistant editors was also appointed to further simplify the editing procedure and attain best results for the readers.

Apart from the editorial board, the designing team has also invested a significant amount of their time in understanding the subject and creating the most relevant covers. They scrutinized every image to scout for the most suitable representation of the subject and create an appropriate cover for the book.

The publishing team has been an ardent support to the editorial, designing and production team. Their endless efforts to recruit the best for this project, has resulted in the accomplishment of this book. They are a veteran in the field of academics and their pool of knowledge is as vast as their experience in printing. Their expertise and guidance has proved useful at every step. Their uncompromising quality standards have made this book an exceptional effort. Their encouragement from time to time has been an inspiration for everyone.

The publisher and the editorial board hope that this book will prove to be a valuable piece of knowledge for students, practitioners and scholars across the globe.

Index

www.ingramcontent.com/pod-product-compliance
Lightning Source LLC
Chambersburg PA
CBHW082100190326
41458CB00010B/3532